An introduction to digital electronics and logic

An introduction to digital electronics and logic

R. H. Joynson

B.Eng., C.Eng., M.I.E.R.E., M.I.E.E.E.
Senior lecturer,
Department of Electronic and Radio Engineering,
Riversdale College of Technology, Liverpool

Edward Arnold

First published 1981
by Edward Arnold (Publishers) Ltd
41 Bedford Square, London WC1B 3DQ

British Library Cataloguing in Publication Data

Joynson, R. H.
  An introduction to digital electronics and logic
  1. Digital electronics   2. Logic design
  I. Title
  621.381′73′7   TK 7868.D5

  ISBN 0-7131-3440-2

Filmset in 10/11 pt Linoterm Times by Keyset Composition
and printed in Great Britain by Richard Clay (The Chaucer Press) Ltd,
Bungay, Suffolk.

Contents

Preface

In just over thirty years, society has been involved in a revolution more important and with more far-reaching effects than the Industrial Revolution.

Since most industries being affected by this 'new technology' do not employ trained electronic technicians, conventionally trained electricians are finding themselves confronted with devices and systems using technology about which they have little knowledge and possibly a fear.

The development of the transistor and integrated-circuit technology has produced a reliable and efficient component available cheaply in large quantities. My aim in this book is to give an understanding of some of the practical aspects of this technology, and to this end I have limited the theoretical content, which I feel is often irrelevant and even confusing to many service technicians 'on the tools'. I am sure that I have not covered everything that would be of interest, or that I would have wished, the constraints of space having limited many points on which I would have liked to expand. For this reason I have chosen to put the emphasis on digital technology.

As well as electricians I hope that the book will be of value to anyone who wishes to become familiar with new developments in these fields. I feel that where possible practical work should be carried out to reinforce the theory, since there is no substitute for seeing for oneself. Those readers needing more detailed information should consult some of the many books available in these fields, some of which are listed at the end of the book.

I should like to acknowledge my debt to all those sources of information which as a teacher one invariably consults and whose source is often forgotten as time passes. Particularly worthy of mention are the many data books and technical manuals produced by the various manufacturers and in particular those companies who have given permission to use copyright material.

Finally I should like to thank those colleagues whose efforts have contributed to the development of the ideas behind this book, in particular Ian Lund and John Crimes, and my sister Sybil Barber for typing the manuscript.

Roy Joynson

Acknowledgements

The author wishes to thank the following organisations for kind permission to reproduce photographs of their equipment: Tektronix UK Ltd (fig. 1.3), Hewlett Packard Ltd (figs 5.2(a) and (b)), Philips Electronic and Associated Industries Ltd (figs 11.1 and 11.4), Mullard Ltd (fig. 11.2), and Standard Pneumatic Motor Co. Ltd (fig. A5.2).

Thanks are also due to the following for permission to make use of copyright material and drawings: Mullard Ltd (fig. 11.3 and the explanation of integrated-circuit manufacture) and Texas Instruments Ltd (figs 13.1, 14.1, and 14.2).

1   Basic electronic test equipment

Introduction

While some routine and breakdown servicing does require frequency counters, digital voltmeters, etc., most testing requirements for general electronics can be covered adequately by just two instruments – a good-quality multimeter and an oscilloscope. The possibilities and weaknesses of these two instruments make up this chapter.

For testing electronic logic systems some additional items are necessary, but this specialised field is the subject of a later chapter.

The multimeter

Multimeters such as the AVO are familiar pieces of equipment to most of us, but there are dangers in their use and certain facilities tend to get forgotten. the following paragraphs are intended as reminders and are written with the AVO model 8 in mind, since it is in such general use. However, the reader should be aware that there are many digital multimeters available which offer many advantages over the conventional moving-coil meter.

Effect of meter resistance

Many good-quality meters are quoted as having a sensitivity of 20000 Ω/V. This means that if the meter is set to the 1 V range the meter resistance will be 20000 Ω. Similarly if set to 10 V the resistance is 200000 Ω, or set to 30 V the resistance is 600000 Ω.

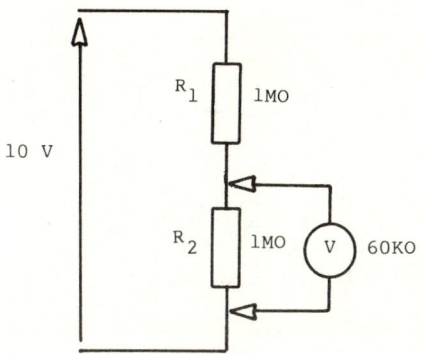

Fig. 1.1 Voltage measurement in a high-resistance circuit

1

These sound like large resistances, but suppose we required to measure the voltage across R_2 in fig. 1.1. This would require the use of the 10 V range, so the meter resistance would be 200 000 Ω (for the AVO model 8). The effect of the meter resistance would be to make the voltage across R_2 fall, from the 5 V it clearly should be to about 1.5 V. The meter would indicate the actual voltage across its terminals, which is 1.5 V (approximately).

The above example shows a disastrous error in an apparently simple measurement. Care must be exercised – especially in making low-voltage measurements in high resistance circuits. In many cases, manufacturers quote the type of meter they have used to obtain test measurements.

Measurement of a.c.

Multimeters generally measure a.c. quantities by using a built-in rectifier and are calibrated to read r.m.s. values. The calibration assumes that the a.c. is a pure sine wave. If the input is not a sine wave, the reading will be in error. In practice, errors are generally small until the distortion of the sine wave is severe. Squarish wave shapes tend to give high readings, and peaky ones give low readings.

Many multimeters are capable of reading a.c. quantities over a wide range of frequencies. For example, the AVO model 8 will read accurately from 25 Hz up to 2 kHz.

Measurement of resistance

Apart from the non-linearities of the resistance scale, the measurement of resistance is quite straightforward. When testing diodes and transistors, however, the resistance measured depends on the direction of the current. It will be found that most multimeters set to the resistance range will drive current out of the black terminal and into the red terminal. This is opposite to what most people expect, so care is required. Figure 1.2 illustrates this point.

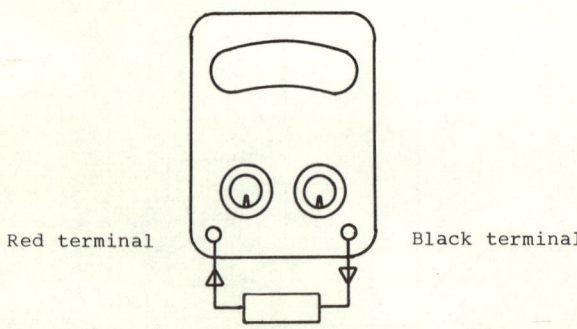

Red terminal   Black terminal

Fig. 1.2 Current direction on a multimeter resistance range

The cathode-ray oscilloscope

The cathode-ray oscilloscope – often referred to as the oscilloscope, 'scope, or CRO – is a most 'powerful' instrument because it allows the engineer to 'see' electrical signals which otherwise would remain a complete mystery. Without this instrument, the development of modern electronics would not have been possible. Without understanding and skill in the use of the oscilloscope, maintenance and repair of such modern electronics becomes haphazard and machine downtime may become extended. A typical oscilloscope is shown in fig. 1.3.

Fig. 1.3 A cathode-ray oscilloscope

What the oscilloscope does

The oscilloscope plots a graph of voltage against time. Instead of using a pencil on a piece of paper, the oscilloscope uses a spot of light on a screen. A graph drawn with a pencil is a record of something that happened in the past, but the oscilloscope plots a graph of what is actually happening as it happens.

A spot of light does not leave a permanent mark as a pencil does, so the graph on the oscilloscope disappears as soon as it is drawn. Fortunately, most of the signals we wish to look at keep on repeating, so we arrange that the graph is being continually redrawn. If the graph is being redrawn fast enough, it appears to our eyes as if it were a permanent graph.

The skill in using an oscilloscope lies in arranging the controls to obtain the fixed display representing the permanent graph and in presenting the most useful graph on the screen.

3

At first, the large number of knobs and switches on the front of an oscilloscope seems baffling. Each control performs a simple function, though, and using an oscilloscope is just a question of starting with the basic functions and gradually building up the complete operating system. Different manufacturers of oscilloscopes provide different controls and facilities. Not all the controls to be described will exist on any one instrument, but alternative names for controls will be indicated in brackets.

Basic controls

The spot of light has various controls associated with it which are described below:

BRIGHTNESS (or INTENSITY)   This allows adjustment of the spot brightness. Note that if the brightness is very high a halo can appear around the spot and there is a danger that the screen may be damaged.

FOCUS   This is adjusted to give the smallest spot size.

ASTIGMATISM   This corrects for optical errors which may cause focusing on vertical lines to differ from horizontal lines. Adjust, *only if necessary*, together with the FOCUS control to give the best focus on vertical and horizontal lines.

The spot of light can be positioned anywhere on (or off) the screen using the following controls:

Y **SHIFT** (*Y* POSITION)   This allows the vertical (or *Y*) position of the spot to be set.

X **SHIFT** (*X* POSITION)   This allows the horizontal (or *X*) position of the spot to be set.

Note that, since the SHIFT controls allow the spot to be set right off the screen, when switching on the instrument it may be necessary to move these controls to bring the spot into view on the screen.

The scale lines (or more correctly the graticule) on the screen are not always easily visible, as they need to be when a measurement is required. On many instruments the scale can be illuminated using the GRATICULE (or SCALE) control, often incorporated in the ON/OFF control.

Vertical (Y) amplifiers

The *Y* direction on the screen is used to plot the signal voltage. In order to allow the oscilloscope to display small signals of a few millivolts or large signals of tens of volts, amplifiers are built into the instrument and are fitted with a gain control. For operating convenience this is a switched control calibrated directly in volts/cm. If this control is set to 2 volts/cm and the display is 3 cm high then the actual signal voltage is 6 V. Some oscilloscopes also have a variable gain control in addition to the switched one. Note that the variable control must always be set to its CAL (i.e. calibrated) position if the volts/cm control graduations are to be used for a measurement.

Many oscilloscopes have two (or even more) vertical amplifiers. These permit two signals to be displayed on the screen at the same time, allowing the signals to be compared. Generally the controls for both amplifiers will be identical and the amplifiers will be identified as CHANNEL 1 and CHANNEL 2 or UPPER TRACE and LOWER TRACE.

Associated with each *Y* amplifier is its input socket. This is coupled to the signal to be displayed using a co-axial cable to minimise noise pick-up (i.e. small unwanted signals). Often the cable ends in a probe, to allow easy connection to the source of the signal. Many probes are marked 10×, which indicates that all voltage readings measured on the screen should be multiplied by ten (the probe introduces a loss of ten).

There is often a three-position input selector marked DC/AC/GND. The DC position is used for most purposes. The AC position will only allow the a.c. part of the signal into the amplifier – all d.c. will be blocked. This position allows detailed examination of small a.c. signals appearing on top of a large d.c. voltage, such as ripple on d.c. power supplies. If the DC position were used, the a.c. signal would be too small to be examined or the display would be off the screen. The GND position puts a ground (or earth) on the amplifier input (without shorting the signal) so that the zero-input position can be located on the screen.

The time-base

This is the name of the circuit which allows the graph to be plotted against time. It drives the spot across the screen from left to right at a controlled speed. The speed is selected on a switched control calibrated directly in ms/cm, μs/cm, or ns/cm. Often there will also be a variable control allowing intermediate speeds to be obtained, but again this control must be set to the CAL position before the calibrations on the switched control are used.

Triggering the time-base

This is one of the most important facilities on the oscilloscope, and its correct use is essential for the effective use of the oscilloscope.

For the display on the screen to appear stationary for a repetitive signal, it is clearly important that the spot retraces the same path each time it moves across the screen. This is the job of the trigger circuit. The trigger ensures that the time-base always starts at a particular point on the waveform.

Various controls are used to select which part of the waveform initiates the time-base. The selector switches are as follows:

±   Selects the positive-going or negative-going edge of the waveform.

NORMAL (AC)/HF/TV LINE/TV FRAME   The NORMAL position is used for all general electronics work, but when it is desired to trigger on a very fast part of the waveform and not on the slower parts the HF position will do this. The two TV positions are clearly specialised facilities for TV servicing.

CH1/CH2/EXT   This selects the waveform from which the trigger point is chosen: either channel 1 or 2, or occasionally it may be useful to trigger on yet another signal fed to the EXT socket.

When the source of the trigger has been chosen by setting the above switches, the trigger is made operative by adjustment of the STABILITY and TRIGGER LEVEL controls. For most makes of oscilloscope, these controls are set as follows:

a) Rotate both controls fully clockwise.
b) Rotate the STABILITY control slowly anticlockwise until the display *just* disappears.
c) Rotate the TRIGGER LEVEL control slowly anticlockwise until a stable trace appears. Note that the d.c. level at which the time-base starts can be chosen by the position of this control.

2   Diodes

What is a diode?

A diode is a device which allows current to flow through it in one direction but not in the other. The direction in which current will flow is called the *forward direction* and the direction in which current will not flow is the *reverse direction*. In practice it is found that a very small current can flow in the reverse direction, but for most uses it is so small that it can be ignored.

Types of diode

There are many types of diode which can be made, including the thermionic diode (valve), the metal rectifier (e.g. the selenium rectifier), and the semiconductor diode. We shall consider only the semiconductor diode, since this is the most significant at present.

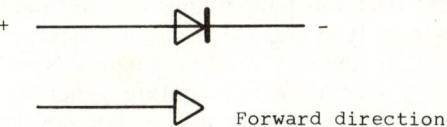

Fig. 2.1 Semiconductor-diode symbol

On a circuit diagram, a semiconductor diode is drawn using the symbol given in fig. 2.1, in which the diode's forward direction is indicated below the symbol. Note how the forward-direction arrow, shown beneath the symbol, forms part of the symbol. Diodes capable of carrying large currents generally have this symbol stamped on them so that the forward direction can be found. Small diodes are generally marked with a painted ring which corresponds with the line in the symbol, as illustrated in fig. 2.2. Occasionally the painted ring may be replaced by a spot of paint (generally red) on older types of diode.

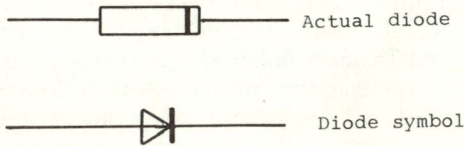

Fig. 2.2 Diode marking related to diode symbol

Many special-purpose devices have been derived from the diode, such as the thyristor (SCR), triac, thyratron, varactor, and zener. Some of these are discussed in chapter 4.

What is a semiconductor?

As its name suggests, a semiconductor is a material whose electrical properties fall between those of the insulators and the conductors. The two most common semiconductor materials for the manufacture of diodes and transistors are silicon and germanium. The addition of small quantities of suitable impurity elements with different numbers of electrons per atom – a process known as 'doping' – produces p-type or n-type semiconductor material.

P-type semiconductor material is produced by doping with impurities which have one less electron per atom than silicon (or germanium). This produces a semiconductor which acts as if it has positive charges or 'holes' which move in the opposite direction to electrons and constitute the current flow.

N-type semiconductor material is produced by doping with impurities which have one more electron per atom than silicon. As electrons are negatively charged particles, this results in an n-type semiconductor. In n-type material, current flow is a normal electron movement.

The electrons in n-type material and the holes in p-type material are *majority* carriers. At room temperature a small number of thermally generated holes in n-type and electrons in p-type semiconductor are present, and these are known as *minority* carriers. Normally these can be ignored; however, increasing temperature generates more minority carriers and eventually results in loss of any special semiconductor properties.

Properties of semiconductor diodes

When p-type and n-type material are combined in a single crystal, the semiconductor diode is formed. As a result of holes and electrons combining, a small region on either side of the junction between the two types of material is created which has no current carriers (apart from a small number of thermally generated minority carriers) and is known as the depletion layer or region.

If a positive potential is applied to the p-type material (or negative to the n-type), the device is said to be 'forward-biased' and it will not allow any significant current to flow below about 0.2V for a germanium diode and about 0.6V for a silicon diode. This corresponds to the potential required to push holes and electrons across the depletion region. At these voltages the current will then flow freely – in other words, the forward diode resistance is very low. If the forward diode resistance is very low then large amounts of current will flow without changing the voltage very much. Thus for a diode conducting current in the forward direction it would be normal to find 0.2 V or 0.6 V across the diode. Note that these figures for the forward voltage are only approximate but give a useful indication when checking diodes in circuit.

If a positive potential is applied to the n-type material (or negative to the p-type), the device is said to be 'reverse-biased' and only a very small leakage current will flow. The reverse bias effectively widens the depletion layer. The reverse current for a germanium diode is significantly greater than that for a silicon diode. Since the current is very small, even at large voltages, it follows that the reverse resistance of a diode is high.

Diode failures

Diodes are generally destroyed by high temperatures, which melt the diode crystal. The high temperatures are generally caused by power dissipated in the diode itself.

In the forward direction, current will generate heat; for example, a silicon diode carrying 30 A would dissipate $30 \times 0.6 = 18$ watts, since its forward voltage would be about 0.6 V. This represents a considerable amount of heat, which is generally removed by bolting the device on to a heat sink. Failure will occur if the heat is not conducted away from the device fast enough. Consequently, when large diodes are replaced it is important that their replacements are securely bolted down (note – maximum tightening torques are quoted by diode manufacturers) using any specified washers, silicon grease (to aid heat flow), etc., and that cooling air to the heat sink is in no way restricted.

Other failures in the forward direction may be due to excessive forward current resulting from breakdown of other components or accidental short circuit. The excessive forward current creates more heat than can be conducted away by the heat sink, and the diode crystal is destroyed by heat. It is worth noting that, since it is very small (see fig. 2.3), the diode crystal heats up extremely rapidly and failures of this type appear instantaneous.

Failures occur in the reverse direction since each type of diode has a limit to the reverse voltage it will withstand. This voltage is generally

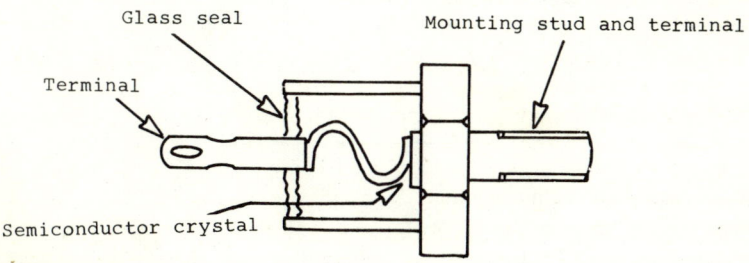

Fig. 2.3 Construction of a large diode

called the 'peak inverse voltage' (PIV). When the PIV is exceeded, the diode crystal allows current to pass. This effect is often referred to as 'breakdown'. The current flowing due to this high voltage creates high temperatures in the crystal, which is rapidly destroyed. Note that it is not the breakdown effect which destroys the diode – it is the temperature.

Transients (voltage or current surges) occur in electrical systems when some disturbance disrupts normal operation of the system. These disturbances are caused by such things as lightning surges, transformers being energised, and load switching, and they may generate voltages and currents which exceed device ratings.

Rectification

One of the most important uses for diodes is in power supplies. It is convenient to distribute electrical power from the generating station to the users as a.c., whereas electronic equipment requires d.c. power. One of the functions of a power supply is, therefore, to convert the incoming a.c. into d.c. This process is called rectification and is performed by a circuit called a rectifier.

The half-wave rectifier
A simple form of the half-wave rectifier is shown in fig. 2.4 and consists merely of a diode placed in series between the a.c. supply and the load. During the first half cycle (marked A in fig. 2.5) of the input sine wave, X is more positive than Y so current is attempting to flow through the diode in the forward direction of the diode and current will flow through the diode and the load. In the forward direction, the resistance of the diode is so low that nearly all the voltage at the a.c. input appears across the load. During the next half cycle (B in fig. 2.5), X is now negative compared to Y so that current is attempting to flow in the reverse direction of the diode and no current will flow, as the diode is now high resistance. If no current – or, to be more precise, only exceedingly small current – flows then there will be no voltage across the load.

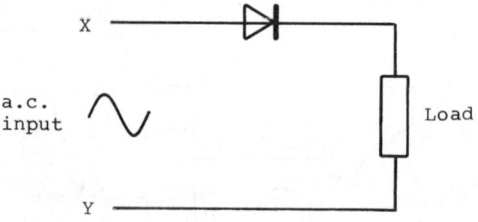

Fig. 2.4 Half-wave rectifier circuit

The voltage appearing across the load is shown in fig. 2.5. The effect of the rest of the input waveform is as just described for A and B, so that C

10

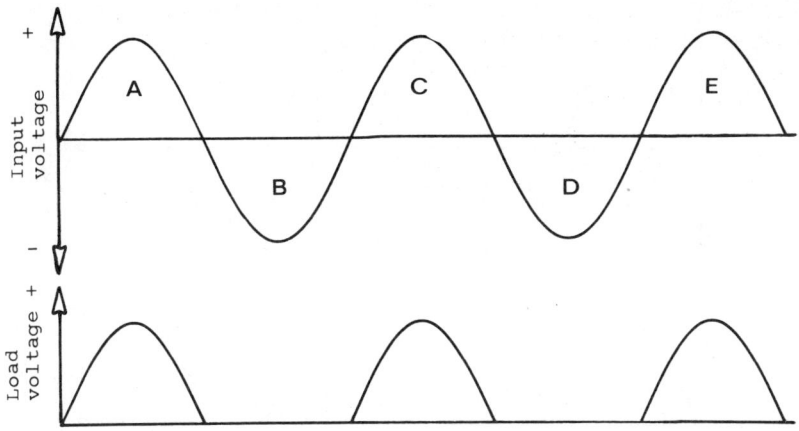

Fig. 2.5 Waveforms for a half-wave rectifier circuit

and E have the same effect as A etc. The voltage across the load consists of half the input waveform, which is where the name 'half-wave rectifier' originates.

The full-wave rectifier

Rectifiers are often fed from a transformer (generally operating on the 240V mains) with a centre-tapped secondary winding which can be used in the full-wave rectifier circuit shown in fig. 2.6.

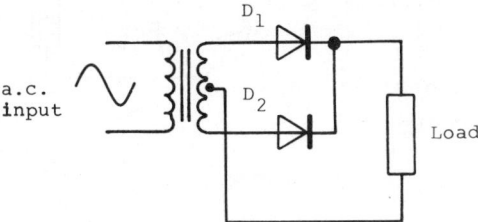

Fig. 2.6 Full-wave rectifier circuit

When the top of the transformer secondary winding is positive during the positive half cycle A in fig. 2.7, current flows from the top half of the winding through D_1 into the load. On the negative half cycle B, the lower end of the transformer will then be positive so that current flows to the load through D_2. Note that the current flows through the load in the same direction irrespective of which diode is conducting – this is undoubtedly d.c.

This circuit is known as a full-wave rectifier, since both halves of the input wave are rectified so as to appear at the output.

11

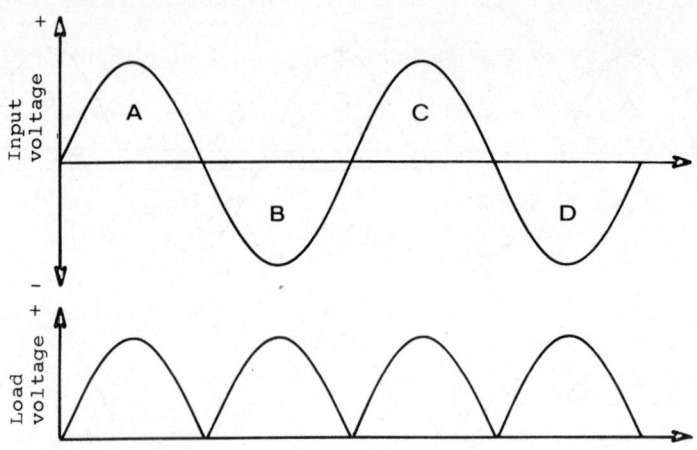

Fig. 2.7 Waveforms for a full-wave rectifier circuit

The bridge rectifier

When a centre-tapped transformer is not available, the bridge rectifier circuit shown in fig. 2.8 can be used to produce the same output waveform.

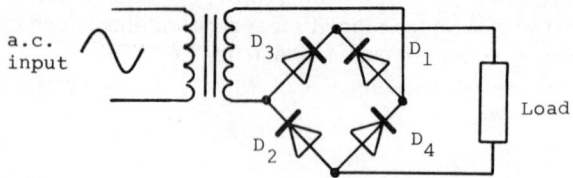

Fig. 2.8 Full-wave bridge rectifier circuit

When the top half of the transformer secondary is positive (A in fig. 2.9), current flows through D_1, through the load from top to bottom, and through D_2. On the negative half cycle (B in fig. 2.9), current flows through D_3, through the load from top to bottom, and through D_4.

While the bridge rectifier can be made up with four separate diodes, it is now common to find the bridge rectifier as one complete package.

Smoothing circuits and voltage stabilisers

While the smoothing circuit is not a diode application, it is included in this chapter to complete the discussion of rectifiers.

The output from the rectifiers discussed above is certainly d.c. (since the current flows through the load in only one direction) but it is not yet

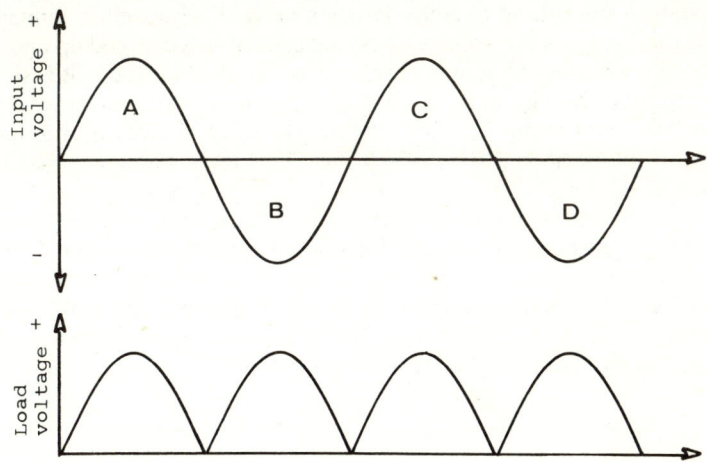

Fig. 2.9 Waveforms for a full-wave bridge rectifier

suitable for driving electric motors or as a power supply for electronic equipment. Electronic equipment particularly requires a smooth d.c., without any of the bumps we have seen in the waveforms so far. The function of the smoothing circuit is to smooth out these bumps.

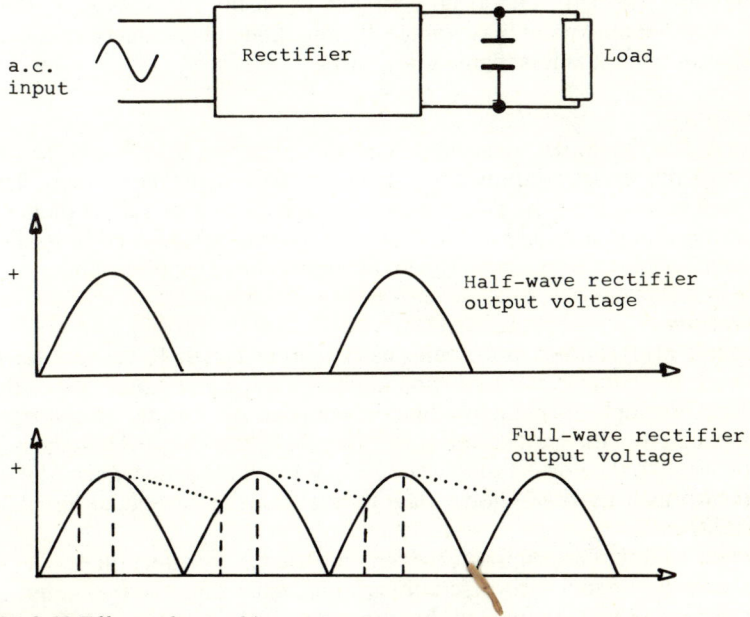

Fig. 2.10 Effects of smoothing

13

A simple smoothing circuit consists merely of a capacitor connected across the load. As the voltage rises, the capacitor is charged up and acts rather like a battery to supply current between the bumps. This is shown for a number of cycles in fig. 2.10, where the original output waveform is shown by a continuous line, the smoothed output waveform by a dotted line, and the periods during which the capacitor is being charged by a dashed line.

The improvement is clearly considerable, and a large enough capacitor provides a good enough output for many types of equipment. Slightly more complicated smoothing circuits are used in some power-supply designs, but generally an electronic circuit called a voltage stabiliser (or sometimes voltage regulator) is used where a large improvement in output quality is needed – refer to chapter 19.

Other diode applications

In addition to being used in rectification circuits, diodes are widely used in electronic circuits to provide a variety of functions, such as clipping, clamping, protection, and detection. Figure 2.11 shows five sample circuits.

Clipping
Figure 2.11(a) shows a simple circuit in which the positive and negative peaks are 'clipped'. Diode D_1 conducts when the input rises approximately 0.6V above the bias voltage E_1, and diode D_2 conducts whenever the input voltage is less than $-(E_2 + 0.6) \text{V}$.

Clamping
Figure 2.11(b) shows a simple circuit in which two diodes are used to 'clamp' the unused inputs of a logic gate to a constant voltage. The voltage drop across the two is approximately 1.2V (for silicon diodes), thus the inputs are held at 3.8V. This also provides protection against voltage spikes or transients driving the inputs above supply voltage.

Protection
Figure 2.11(c) shows a diode being used to protect an SCR (see chapter 4) which is switching current into an inductive load, e.g. a motor. When the current through an inductive load is switched off rapidly, the voltage induced across the load can be very large. Since this induced voltage is in opposition to the normal supply (Lenz's law), a diode connected as shown – known as a flywheel diode – allows the induced current to circulate harmlessly.

Figure 2.11(d) also shows a diode being used as a protection device, in this case to prevent a reverse-voltage spike exceeding the base–emitter reverse breakdown voltage of the transistor.

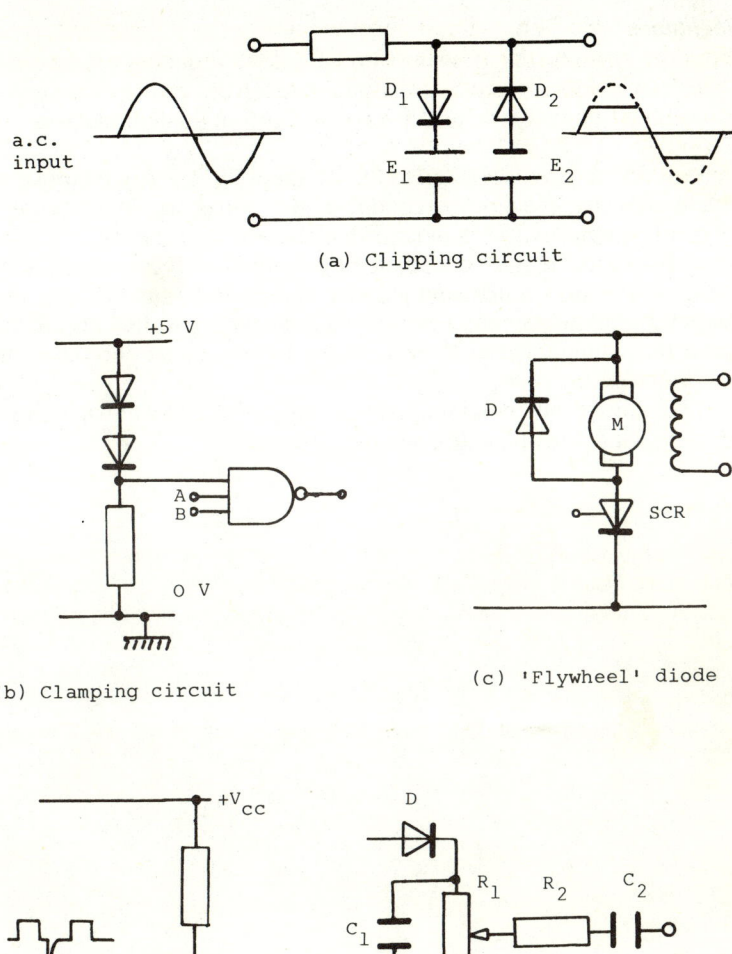

(a) Clipping circuit

(b) Clamping circuit

(c) 'Flywheel' diode

(d) Transistor base
protection

(e) Demodulation circuit

Fig. 2.11 Some diode applications

Demodulation

Modulation permits the transmission of a low-frequency signal on a high-frequency carrier wave. In a radio receiver, the low-frequency information has to be extracted by a process known as 'demodulation' or 'detection'.

If a modulated wave is applied to the circuit of fig. 2.11(e), the positive half cycle will cause the diode to conduct and C_1 will charge up to the peak value of the carrier wave, provided that the time constant $\tau = C_1R_1$ is long compared to the period of the carrier, which can be assumed to be constant. If τ is also much smaller than the periodic time of the low-frequency modulating signal, C_1 will be able to charge and discharge at this rate and the voltage across R_1 will closely follow the amplitude of the modulating wave.

R_2 will eliminate any remaining carrier signal, and finally C_2 will remove any d.c., leaving only the audio-frequency signal.

3 Transistors

What is a transistor?

Apart from the obvious fact that a transistor is not a small type of radio, it is often difficult to appreciate just what it is and does. We seem to be bombarded with the complicated ideas on transistors – of which there are a great many – and miss out on the simple one which follows:

> a transistor allows a current to flow through it, but the size of this current is controlled by another very small current.

Symbols and names used for transistors

Like the semiconductor diode, the transistor is made from a crystal of either germanium or silicon to which other 'impurity' materials are added to give the desired properties. Different types of impurity can be added to give two basic types of transistor, which are referred to as npn and pnp. We shall deal with the npn type almost exclusively, since this is the most common type in modern equipment.

The npn transistor symbol is contained inside the circle in fig. 3.1. (Note that the symbol may be drawn either with or without the circle.) The three electrodes – collector, base, and emitter – are also shown on the diagram. The normal current directions are also indicated. The currents are generally referred to as I_c, I_b, and I_e, where these are currents flowing into the collector, into the base, and out of the emitter respectively.*

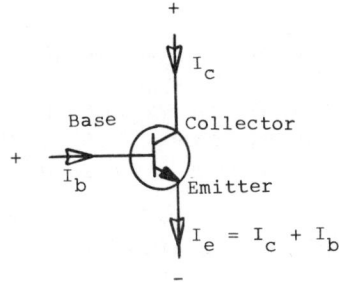

Fig. 3.1 Symbol for the npn transistor

*Lower-case subscripts have been used throughout the book, to avoid unnecessary confusion, but readers should be aware that more advanced texts use capital subscripts to indicate d.c. conditions and lower-case for a.c. conditions.

Note that the arrow on the emitter part of the symbol indicates the current direction in the emitter.

Since transistors depend on *both* holes and *and* electrons for their operation, they are referred to as 'bipolar' devices.

Basic transistor action

It has already been said that the transistor is a device which enables a small current to control a much larger one. This is illustrated in fig. 3.2, where the main current flows through the transistor from the collector to the emitter. This main current is controlled by a much smaller current which flows into the base and out of the emitter. It should be noted at this stage that $I_e = I_b + I_c$ but that, since I_b is very small, for most practical purposes I_e can be taken as being the same as I_c.

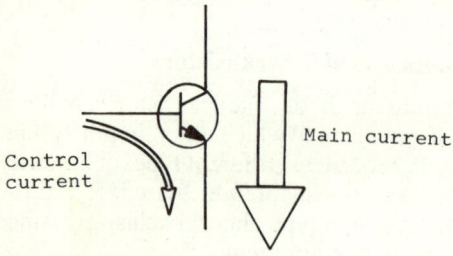

Fig. 3.2 Basic transistor action

For a particular transistor it is found that under normal conditions there is a fixed ratio between the currents I_c and I_b, known as the *current gain* (symbol h_{fe}). The value of h_{fe} can be found on manufacturers' data sheets. The mathematical representation is

$$I_c = h_{fe}I_b$$

The value of h_{fe} for a transistor can be from about 10 up to about 1000, depending on the type.

The symbol β ('beta') may be used instead of h_{fe} for current gain, particularly in older books. However, h_{fe} is the symbol used in manufacturers' data books and is to be preferred.

Relationship between diodes and transistors

A transistor can be considered as two diodes back to back, as one of the original symbols for the transistor indicates (fig. 3.3).

The base current (control current) is now seen to flow through one of these diodes in the forward direction, so if the transistor were made from silicon then the voltage drop between base and emitter would be expected to be 0.6V approximately. This voltage is generally abbreviated to V_{be},

18

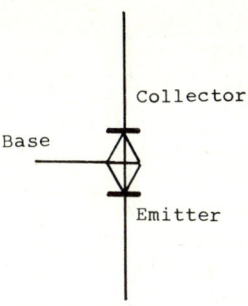

Fig. 3.3 Original symbol for the npn transistor

and a silicon transistor will be found to have $V_{be} \simeq 0.6\,V$ when base current is flowing.

Testing transistors

This similarity to two diodes enables a simple test to be devised using an ohmmeter. It can be seen that between the base and the emitter and the base and the collector we have a simple pn junction. Thus if the base–emitter resistance is taken with the meter leads both ways round, one reading should be very high and the other low. Similar readings should be obtained from base to collector. However, between the emitter and the collector there are two diodes and it can be seen that one pn junction will be reverse-biased whichever way round the ohmmeter is connected, thus both readings should be high.

Thus by taking six simple resistance measurements it is possible to obtain a very good indication of whether a transistor is likely to be serviceable – certainly, if it does not meet the above requirements it is faulty.

While with some transistors and a thorough understanding of the biasing requirements it is possible to get an indication of transistor current gain and so determine whether the device is serviceable, in practice the majority of people seem quickly to forget the method and the above technique is generally adequate.

The method can be used to identify unmarked transistors and the individual leads, but in practice most people involved in repair of electronic equipment should know whether the device is npn or pnp and have data sheets to identify pin layouts.

The transistor in a simple circuit

The transistor circuit shown in fig 3.4 uses a voltage supply of 10 V, typical for transistor circuits. Supply voltages may range from as little as 1.5 V up to 1000 V but generally lie between 5 V and 30 V. The resistors in a

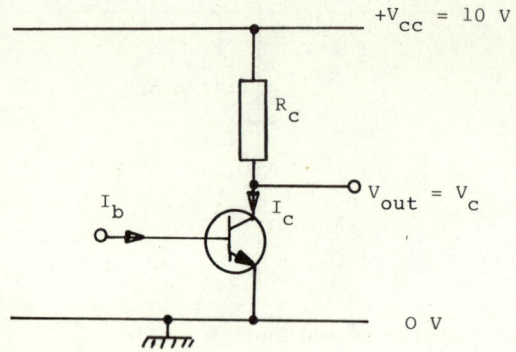

Fig. 3.4 A simple transistor circuit (1)

transistor circuit are often identified with the electrode to which they are connected, so R_c is the collector resistor. (A resistor in this particular position may also be called a load resistor.)

This circuit operates in the following way. The control current I_b fixes the value of I_c. This current flows through R_c and sets up a voltage across R_c which gives the output voltage $V_{out} = V_c$.

This can be stated more clearly in a mathematical form:

Input current $= I_b$ amperes

Collector current $= I_c = h_{fe}I_b$ amperes

Voltage across $R_c = I_cR_c$ volts

Output voltage $= V_{out} = V_{cc} - I_cR_c = (10 - I_cR_c)$ volts

It should be noted that while the effect of increasing I_b is to increase I_c, it will *decrease* the output voltage V_{out}. This is an important point. We now have an input current controlling the output current and the output voltage.

This method of starting at the transistor input and working through to the output is the standard approach to finding out what the various conditions in a transistor circuit should be. The results of these simple calculations can then be compared with measurements on the actual circuit in order to locate a fault in a logical manner.

Let us use this method to find the output voltage for the circuit in fig. 3.5. First we need to calculate the value of I_b, which we obtain as follows:

$V_{be} = 0.6$ volts

Voltage across $R_b = V_{cc} - V_{be} = 20\,\text{V} - 0.6\,\text{V}$

$\simeq 20$ volts

20

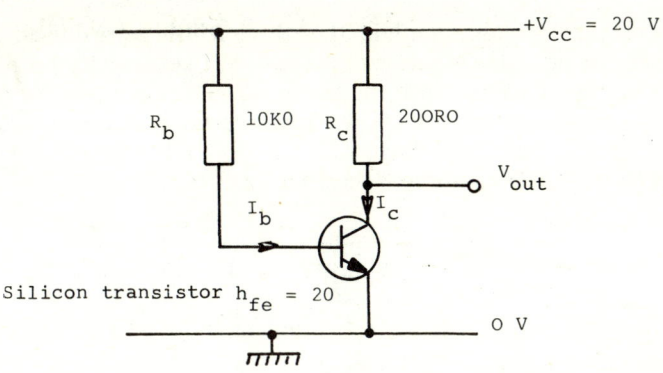

Fig. 3.5 A simple transistor circuit (2)

$$I_b = \frac{\text{voltage across } R_b}{10000} \text{ amperes}$$

$$\simeq \frac{20}{10000} \text{ amperes}$$

$$= 2\,mA$$

$$I_c = h_{fe} I_b$$

$$= 20 \times 2\,mA$$

$$= 40\,mA = 0.040 \text{ amperes}$$

Voltage across $R_c = 200 \times 0.040$ volts

$$= 8 \text{ volts}$$

$$V_{out} = 20\,V - 8\,V$$

$$= 12 \text{ volts}$$

Transistor amplifiers

What is an amplifier?
An amplifier is an electronic circuit which when given a signal at its input reproduces that signal increased in size at its output.

A simple amplifier
Suppose we require an amplifier to accept a current signal at its input and produce a current output. This is achieved merely by using a transistor. I_b is the input signal and I_c is the output. The gain of the amplifier is the ratio of output to input which in this case is h_{fe}.

21

If we require a voltage amplifier then this is a little more difficult, since the transistor operates in terms of current. We get around this by converting the input voltage into the current I_b and converting the output current I_c into the output voltage. These conversions are readily achieved using resistors, since the voltage across a resistor is proportional to the current through it. The circuit is shown in fig. 3.6.

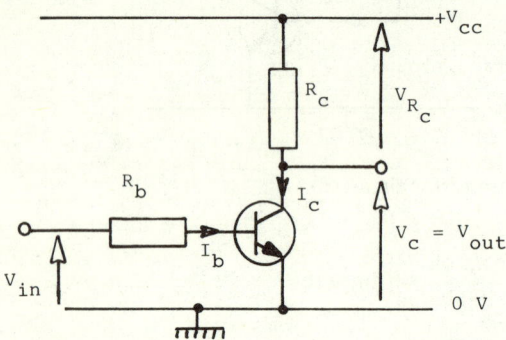

Fig. 3.6 A simple voltage amplifier

As I_b rises, V_{out} falls since $V_{out} = V_{cc} - I_c R_c$; i.e. the supply voltage V_{cc} is fixed and $I_c R_c = h_{fe} I_b R_c$ is increasing. Since small changes in I_b cause relatively large changes in I_c, a relatively larger change in the voltage across R_c occurs than that required to cause the original change in I_b, thus the circuit produces amplification, though the output is 'upside down' or phase-inverted compared to the input. For most purposes this is not a problem.

This type of circuit – known as the common-emitter amplifier – does however have problems which result from changes in the temperature of the transistor. If the temperature increases then, due to a side effect, I_c will increase; also V_{be} falls, causing I_b to increase which in turn increases I_c. The output voltage is therefore unstable and the circuit is not satisfactory.

For this reason, circuits for amplifying d.c. signals are more complicated in order to overcome this problem and will be considered later. Fortunately, circuits for amplifying a.c. signals can be built to work satisfactorily using just a few more components, but to understand these circuits we must first digress slightly and consider the emitter-follower circuit.

The emitter-follower amplifier
This is a very simple circuit (fig. 3.7) in which the action of the transistor causes V_{out} to follow V_{in} – hence the name. For the transistor to operate, the base–emitter junction must be operating in the forward direction so

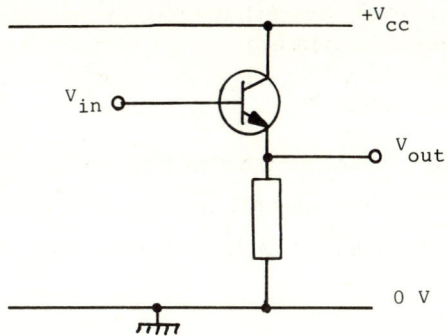

Fig. 3.7 The emitter follower

that $V_{be} = 0.6\,V$ for a silicon transistor. The result is that V_{out} follows V_{in} with a small difference of about $0.6\,V$, and the gain is therefore 1.

The temperature-stabilised a.c. amplifier

The emitter-follower circuit has many uses in its own right, generally because it has a high input resistance and a low output resistance. Our main interest, however, is in its use in the a.c. amplifier circuit.

The circuit diagram is shown in fig. 3.8. The capacitors on the input and the output are used to ensure that only a.c. is fed into or out of the circuit. The circuit can be considered as d.c. levels with the a.c. superimposed on top.

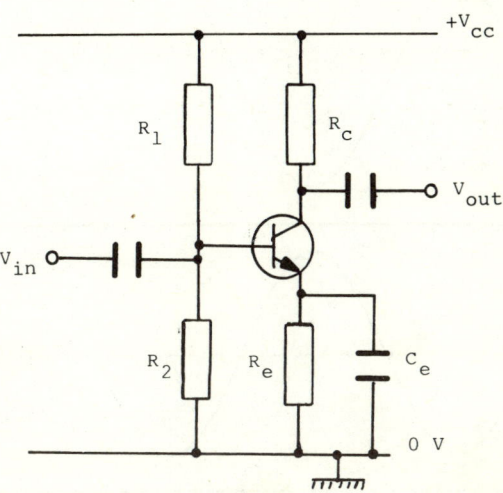

Fig. 3.8 A temperature-stabilised a.c. amplifier

Considering d.c. conditions first, the resistor chain R_1 and R_2 sets the voltage on the base of the transistor:

$$\text{transistor base voltage } V_b = \frac{V_{cc} \times R_2}{R_1 + R_2}$$

The emitter voltage can now be found as in the emitter follower:

$$V_e = V_b - 0.6\text{V} \quad \text{for a silicon transistor}$$

The emitter current is approximately equal to the collector current, so

$$I_c \simeq I_e = \frac{V_e}{R_e}$$

Finally $\quad V_c = V_{cc} - I_c R_c$

We have thus obtained voltages in all parts of the circuit, so measurements with a multimeter on an actual circuit could quickly find out if anything was wrong.

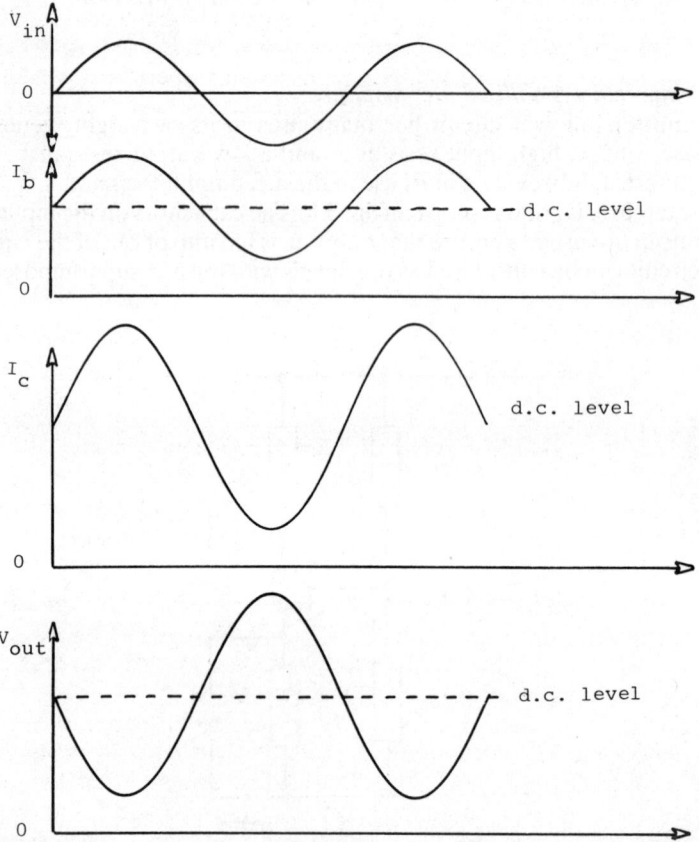

Fig. 3.9 Waveforms for the a.c. amplifier

24

The effect of introducing an a.c. input signal V_{in} will be to modify the base current. This is illustrated in fig. 3.9 for a sine-wave input.

Again it should be noted that the input is inverted (upside down) in comparison with V_{in}. V_{out} has the a.c. output signal superimposed on a d.c. level – if necessary, the d.c. level can be removed by means of a capacitor.

The effects of R_e and C_e in this circuit are important. R_e is included since it makes the transistor behave as an emitter follower. One of the advantages of the emitter follower is that it does not suffer badly from changes in temperature. The gain of the emitter follower presents a problem, so its effect for a.c. signals is eliminated by C_e which bypasses the resistor R_e for the a.c. signals. The addition of C_e does not affect the temperature-stabilising properties of the emitter follower, which operates on d.c.

Switching circuits

Figure 3.10 gives the circuit of a transistor used for switching. This circuit not only resembles the simple voltage amplifier of fig. 3.6 but is identical to it – the only difference is in the way in which it is used. A switching circuit is concerned only with the extremes of operation, unlike the amplifier which is concerned with what happens in between the extremes.

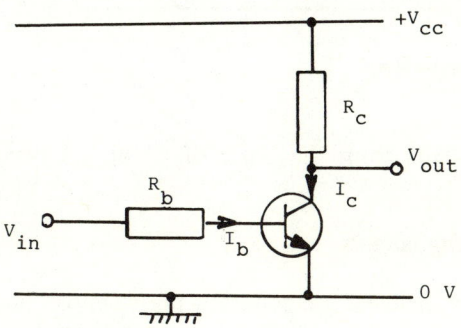

Fig. 3.10 A simple switching circuit

When V_{in} is made zero, it follows that I_b will be zero. If I_b is zero then so will I_c be zero. The voltage across R_c will also be zero, so V_{out} will be equal to V_{cc}. This is clearly one of the extreme conditions, and the transistor which is carrying no current is said to be *cut-off*.

The other extreme is when V_{in} is made a large value. Let us consider what happens to V_{out} as we gradually increase V_{in}. As V_{in} rises so do I_b, I_c, and the voltage across R_c, so V_{out} will fall. What happens when the voltage across R_c becomes almost the same as V_{cc}? If we continued to increase V_{in} we could hardly expect to make V_{out} negative. In fact, V_{out} will stay at a voltage which is almost zero even though we continue to increase V_{in}. In

25

this condition the transistor is said to be saturated (or bottomed). The small voltage which remains across the saturated transistor is called the saturation voltage, $V_{ce(sat.)}$, and is typically 0.1 V.

If we arrange that the input to this simple circuit is either sufficiently low to cause cut-off or sufficiently high to cause saturation then the action of the transistor is similar to a switch controlled by an input voltage (or current). This is illustrated in fig. 3.11. The switch is open (cut-off) when V_{in} is very low or zero, and the switch is closed (saturated) when V_{in} is very high. At this point it should be noted that the terms 'on' and 'off' are often used for the saturated and cut-off conditions.

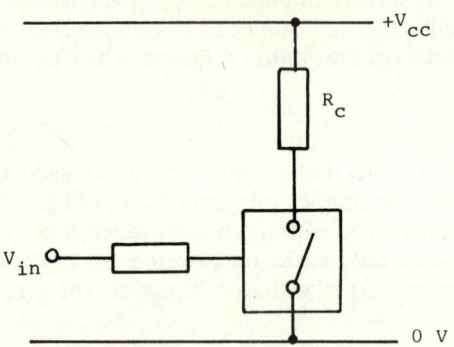

Fig. 3.11 Switching function

Switching circuits form the basic of all digital systems, including computers.

Practical switching circuits

Silicon transistor switches are generally fairly simple, and the circuit in fig. 3.10 is often adequate. Note that, because this circuit is operated at its extremes, as long as V_{in} is low enough to ensure cut-off or high enough to provide plenty of base current to ensure saturation the effect of temperature change is negligible.

In practice it is found that the small saturation voltage of a transistor switch can be connected directly to the input of a second similar switch and ensure cut-off in the second one.

Germanium transistor switches are more difficult to turn off, so a second supply-voltage rail often is introduced to ease this problem. This is shown in fig. 3.12, where the negative supply voltage operating through resistor R ensures that when V_{in} is at zero, or the saturation voltage from a previous switch, the base current will be zero and the germanium transistor will be cut-off.

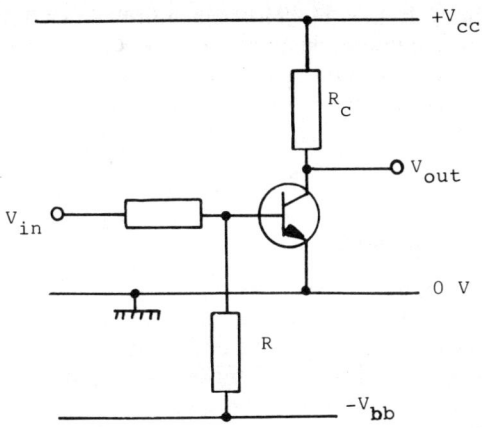

Fig. 3.12 A germanium-transistor switch

This is necessary since V_{be} for a germanium transistor is so much less than V_{be} for a silicon transistor.

Speed-up capacitors are used in some circuits where the speed with which the transistor moves from cut-off to saturation or vice versa is important (fig. 3.13).

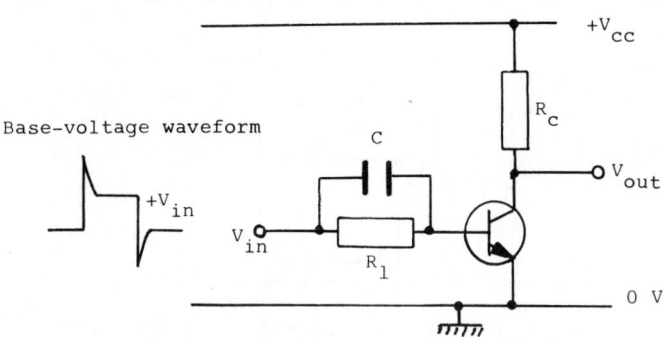

Fig. 3.13 A switching circuit with a 'speed-up' capacitor

The speed-up capacitor provides initial and end 'spikes' to the switching voltage. It achieves this because the charge between its plates cannot change instantaneously. If an input pulse is applied without the capacitor, the resistor R_1 will limit the rise of voltage at the base. If capacitor C_1 is added and the input pulse is applied, the left-hand plate of C_1 is immediately raised to $+V_{in}$ and, since the capacitor cannot change its charge instantaneously, it must follow that the right-hand plate is also raised instantaneously to $+V_{in}$. The capacitor then charges up and the base voltage falls. However, the initial spike ensures that the transistor is overdriven initially. A similar reasoning explains the negative spike on removal of the input pulse.

Direct-coupled amplifiers

D.C. and slowly varying a.c. signals cannot be amplified by conventional amplifiers using capacitive coupling, due to the capacitor reactance. Amplifiers used for amplifying these low-frequency and d.c. signals are called 'direct-coupled' or just 'd.c.' amplifiers. One simple, but common, type is described below. The operational amplifier described in chapter 13 is also suitable for use as a d.c. amplifier.

The Darlington-pair transistor
Figure 3.14 shows the basic circuit using two transistors – these can be two individual transistors or a composite specially constructed device.

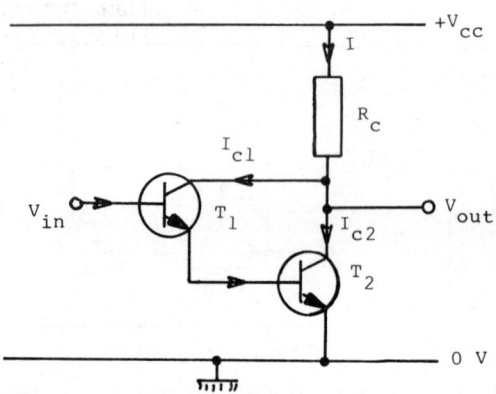

Fig. 3.14 Darlington-connected transistors

This arrangement has a high input resistance and a very high current gain which can be shown to be approximately $h_{fe1} \times h_{fe2}$, where h_{fe1} and h_{fe2} are the gains of T_1 and T_2 respectively.

In practice, more sophisticated circuits may be used – particularly if good thermal stability is required.

PNP transistors

The pnp transistor is essentially the opposite of the npn transistor. The symbol for the pnp transistor is given in fig 3.15. The only difference to the npn symbol is that the arrow on the emitter points inwards. This arrow indicates the direction of current flow in the emitter and implies that all currents flow in the opposite direction to those for the npn device. Power supplies must also be of the opposite polarity.

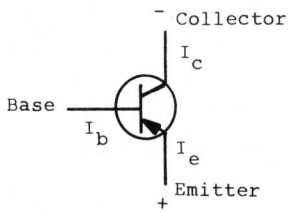

Fig. 3.15 Symbol for the pnp transistor

In circuits using only pnp transistors, it is conventional to draw a negative power-supply rail at the top of the circuit diagram. This is illustrated in fig. 3.16. Often circuits contain both pnp and npn transistors, and in this case it is conventional to show a positive power-supply rail at the top – see fig. 3.17.

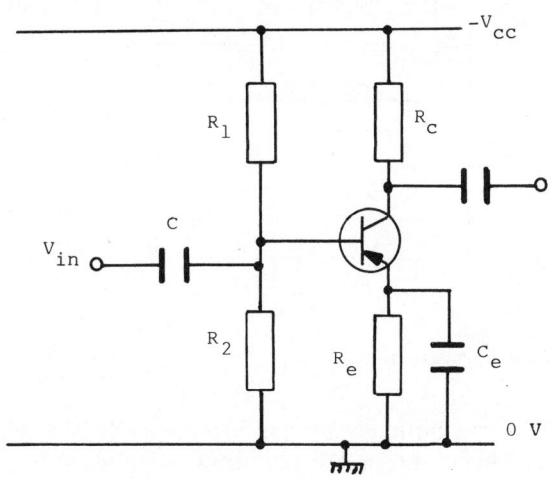

Fig. 3.16 A simple amplifier using a pnp transistor

29

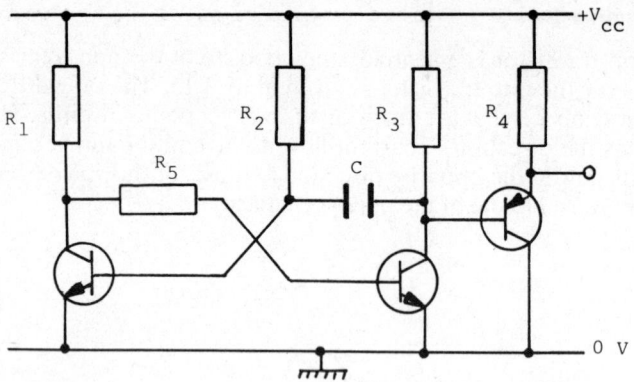

Fig. 3.17 A circuit using npn and pnp transistors

The junction-gate field-effect transistor (JUGFET)

Compared to the bipolar transistor, field-effect transistors (FET's) have higher input impedance ($\simeq 10^6\,\Omega$), can be less noisy, and are thermally more stable, but they are not able to operate at such high frequencies. They have three terminals – called drain, source, and gate – as shown in fig. 3.18.

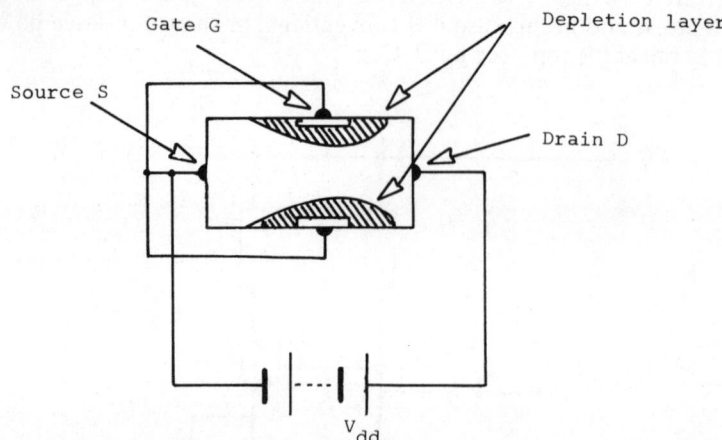

Fig. 3.18 JUGFET construction

A bar of silicon, normally n-type but p-type may be used, forms a 'channel' into which a p-type gate is diffused, resulting in depletion layers around the junctions. Connection of a battery V_{dd} as shown widens the depletion layers at the drain end, because at this end of the bar the channel-to-gate diodes are reverse-biased due to the voltage drop along the bar.

For small values of V_{ds} a current I_d of electrons (majority carriers) proportional to V_{ds} flows through the channel. However, at a certain value of V_{ds} known as the 'pinch-off' voltage V_p, the effective channel width is zero and further increase in V_{ds} has little effect on I_d. This is shown in fig. 3.19.

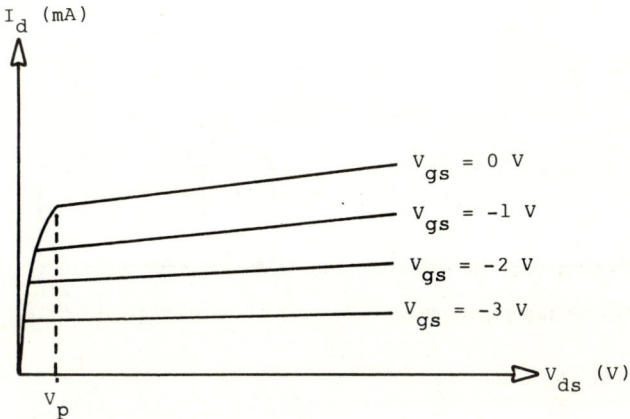

Fig. 3.19 JUGFET output characteristics

If a battery is connected to make V_{gs} negative (gate negative with respect to source) the gate–channel junction will be reverse-biased and the depletion layer is widened. This has the effect of reducing the pinch-off voltage.

The above device is an n-channel JUGFET. The p-channel JUGFET has an n-type gate and reversed voltage and current directions. The symbols are shown in fig. 3.20.

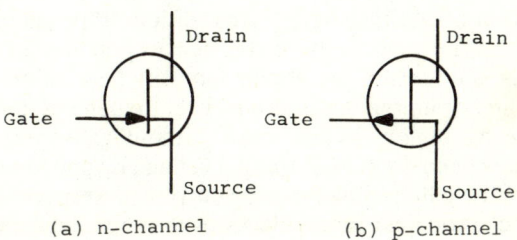

Fig. 3.20 JUGFET symbols

Metal–oxide field-effect transistors (MOSFET's)

One technique used to make FET's is known as metal–oxide–semiconductor technology and uses a structure in which an electric field is created in a metal gate, an oxide insulating layer, and the semiconductor channel. This technique has its major application in the construction of integrated circuits, particularly memories (see chapter 11).

The basic construction of the two types of MOSFET – the enhancement and depletion types – is shown in fig. 3.21.

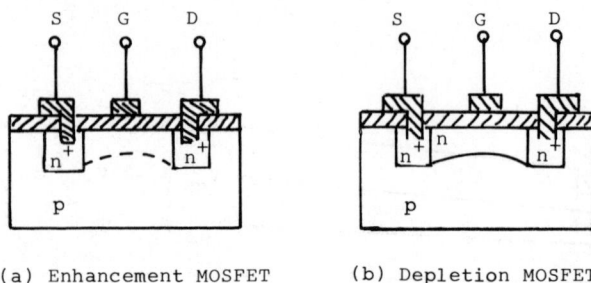

(a) Enhancement MOSFET   (b) Depletion MOSFET

Fig. 3.21 MOSFET construction

In both types of MOSFET the gate is isolated from the substrate by a very thin silicon-dioxide insulating layer (because of this, these devices are sometimes referred to as insulated-gate field-effect transistors or IGFET's), thus very little gate current can flow.

For the enhancement MOSFET, fig. 3.21(a), a drain–source voltage without bias does not result in a drain current since one pn junction will be reverse-biased. If a potential is now applied between the gate and the source (gate negative), a negative charge will be induced in the silicon dioxide closest to the terminal. This will induce a positive charge in the silicon dioxide next to the substrate, which will in turn cause electrons to build up close to the surface, forming a negative channel between the heavily doped n^+ regions permitting current to flow easily between drain and source.

The depletion MOSFET, fig. 3.21(b), differs from the enhancement MOSFET in that the region between the heavily doped source (i.e. where relatively large amounts of the doping impurity have been added to the pure silicon) and drain regions is also n-type, though not quite as heavily doped as the n^+ regions. Thus, even without gate-to-source bias, a current will flow from drain to source if a voltage is applied between drain and source. If the gate is now made negative with respect to the source, the resultant electric field forces electrons out of the channel, depleting the area of charge carriers and reducing current flow. This device can, however, also be operated in the enhancement mode.

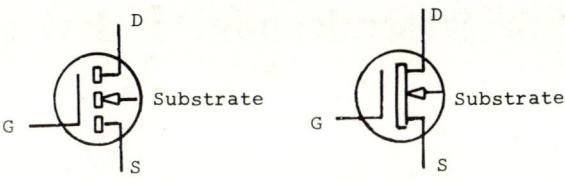

Fig. 3.22 MOSFET symbols

Figure 3.22 shows the symbols for the two types of MOSFET.

Since FET's rely only on electron flow (or hole flow for p-channel devices) in the channel, they are unipolar devices.

Comparison of FET's and bipolar transistors

There is considerable overlap in the suitability of JUGFET's, MOSFET's, and bipolar transistors for specific applications, e.g. all three types can be used as switches. However, there are distinct areas in which each type offers particular advantages, and the most important differences stem from the different input impedances.

The choice of device for a particular application depends on the source resistance – the device resistance must 'match' the source to minimise drift (change in current or voltage due to temperature), noise (random unwanted voltage), and shunting (the short-circuiting effect of other circuits in parallel). For source resistances below a few thousand ohms, bipolar transistors are best; above this range to about 10MΩ JUGFET's are preferred; above 10MΩ MOSFET's are most suitable.

MOSFET's and JUGFET's make very good switches but are slower than bipolar switches. The small size and ease of construction combined with low power consumption make the MOSFET particularly suitable for the manufacture of complex digital circuits, particularly memories.

Due to its ability to provide relatively high voltage gains, the bipolar transistor tends to be the general-purpose amplifying device.

4 Special semiconductor devices

Zener diodes

A zener diode is a special diode which makes use of the breakdown effect when the diode is operated in the reverse direction. The breakdown occurs at a specific voltage, and current starts to flow in the reverse direction. This breakdown voltage remains constant even though the reverse current alters, so the zener diode is used as a voltage stabiliser. The device symbol is shown in fig. 4.1.

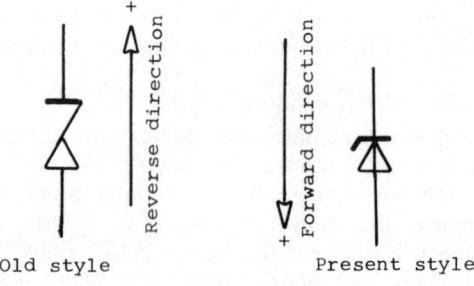

Fig. 4.1 Zener-diode symbols

A wide range of breakdown voltages is available in zener diodes, typically from about 3 V up to 100 V or more. This gives the devices a wide range of applications. Since the voltages across a zener are much greater than a normal diode's forward voltage, the power dissipated in the device can be large for relatively small currents, and heat sinks may be necessary.

The zener as a voltage stabiliser

Probably the most common application for the zener is in providing defined voltages for power supplies.

In the circuit of fig. 4.2, which uses a zener with a 10 V breakdown, the output voltage V_{out} will be 10 V. The output current I_{out} can vary (within limits) without V_{out} changing significantly from 10 V. Also, the input voltage V_{in} can change without any alteration in V_{out} (although clearly V_{in} must always be somewhat greater than 10 V and not so great that the heat-dissipation limit in the zener is exceeded). Summarising, V_{out}

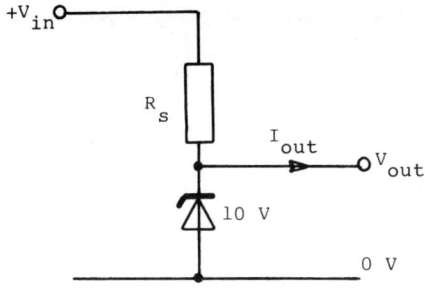

Fig. 4.2 A simple zener voltage stabiliser

remains constant at 10V even if V_{in} and I_{out} change; i.e. the output voltage is stable.

The zener in stabilised power supplies

A simple power supply with a zener-stabilised output is shown in fig. 4.3. The smoothed rectified output at X is shown in the figure. (The half sine waves are indicated to show how the waveform arises.) As long as the lowest parts of the waveform at X are well above the zener voltage, the output will remain a pure d.c. level which, within limits, will not alter with changes in output current or changes in a.c. input voltage.

In practice, the simple circuit in fig. 4.3 will not allow large output-current variations without the use of zeners dissipating large amounts of

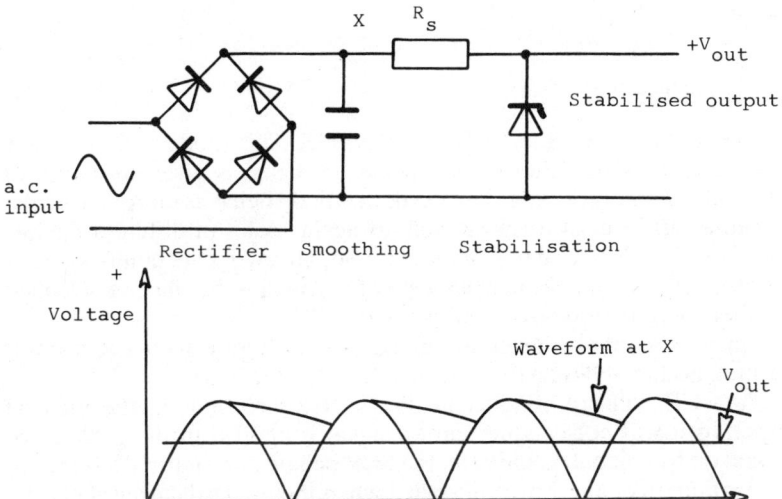

Fig. 4.3 A power supply with zener-stabilised output

power. It is often more convenient to use transistors in addition to the zener. With this type of circuit, the transistor dissipates virtually all the power, which will be much less than for the equivalent circuit using just a zener.

A simple circuit using a transistor is shown in fig. 4.4. In this circuit, the resistor R does not carry the main output current and serves only to provide current to the zener and the base current for the transistor. The main current flows through the transistor from collector to emitter. The output from this circuit will be driving a load, and the transistor can be considered as an emitter follower. It follows that the output voltage will be approximately $0.6\,V$ (for a silicon transistor) below the zener voltage. A capacitor C_2 is placed across the zener to reduce the output noise, since zeners are noisy devices.

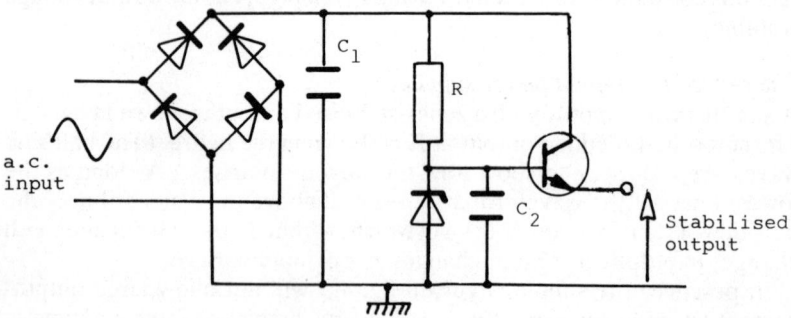

Fig. 4.4 A simple transistor-stabilised power supply

The thyristor

The thyristor is yet another form of diode. This diode will not allow current to flow in either direction until a signal is given to a control electrode. After receiving this signal, the diode behaves more or less like a normal diode until it is switched off again. A lot of different devices based on this idea are being manufactured, all having the family name of thyristor. We shall consider one type of thyristor – the silicon controlled rectifier (SCR) – and how it can be used.

The SCR symbol is given in fig. 4.5. The symbol is the same as for a normal diode but has an extra electrode called the gate.

The SCR will not conduct in the reverse direction. In the forward direction it will not conduct until a signal is applied to the gate. Once turned on by a signal on the gate, the SCR behaves as a normal diode until it is turned off again. Turn-off occurs when the forward current falls to a level below the 'holding current' (a quantity specified by the manufacturer). It is not possible to turn off an SCR using the gate.

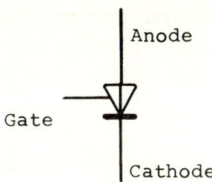

Fig. 4.5 The silicon-controlled-rectifier symbol

The signal required to turn on an SCR is just a direct current flowing into the gate. This current must be above a minimum value specified by the SCR manufacturer. Closing the contacts in fig. 4.6 would supply the small gate current to turn on the SCR so that current flows in the load. The load current will continue to flow after the contacts have been opened again.

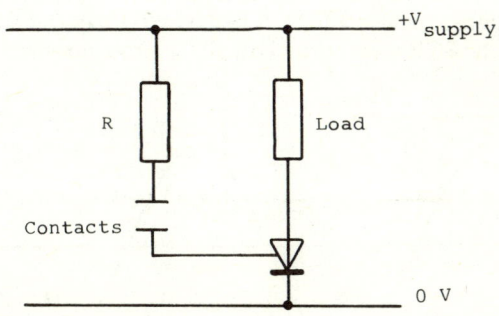

Fig. 4.6 A simple trigger circuit

Load current will continue to flow until it falls below the holding current, e.g. by increasing the load resistance or reducing V_{supply}.

In more sophisticated equipment it is normal to fire (turn on) the SCR by a series of pulses on the gate. 'Burst firing', as it is called, is a more reliable method of control than the simple direct-current method.

Uses of the SCR

The SCR has many important uses in the industrial field. One of the most important is to control the speed (or torque) of an electric motor either for a machine drive or for transport such as the fork-lift truck. When a motor is to be driven from the a.c. mains using SCR's, a d.c. motor is employed and the mains is rectified in a controlled fashion to give the variation in speed. The control method generally employed is called phase control. Phase control is used below in the simple lamp-dimmer circuit.

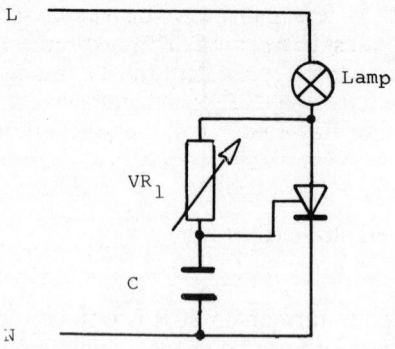

Fig. 4.7 A simple lamp-dimmer circuit

If we ignore VR_1 and C, the lamp-dimmer circuit shown in fig. 4.7 is similar to a half-wave rectifier. If the SCR were turned on all the time, the lamp would be at roughly half brightness since it would be receiving only half the a.c. mains, because the SCR behaves like a normal diode. On the other hand, if the SCR were turned off all the time, no current would flow

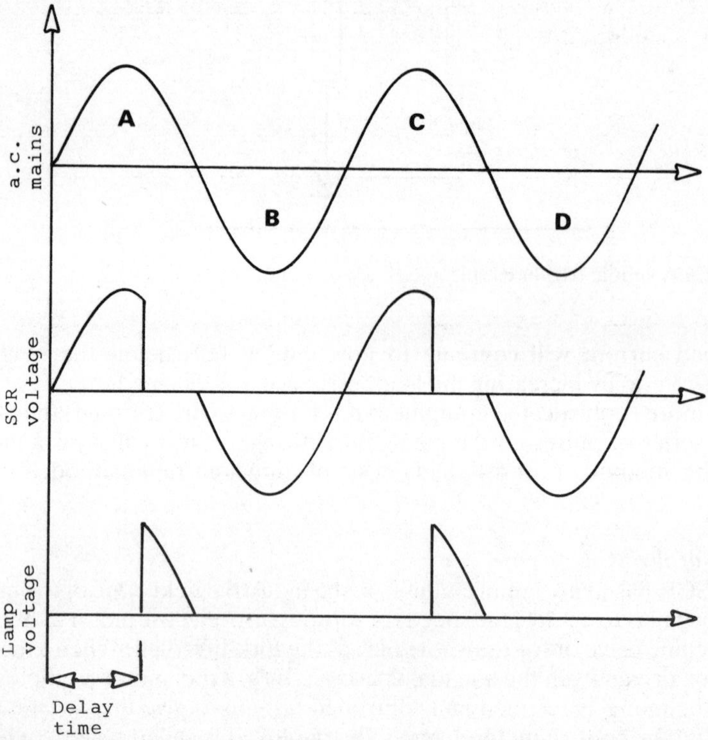

Fig. 4.8 Waveforms for the circuit of fig. 4.7

and the lamp would be extinguished. This simple circuit uses VR_1 to provide a lamp brightness between these two extremes.

If we consider the first half cycle A of the a.c. mains waveform in fig. 4.8, initially C is discharged so there is no gate current. The SCR is thus turned off. The voltage on the SCR will then rise with the rising edge of the mains waveform. As the SCR voltage rises, current flows through VR_1 into C and the gate. As C charges up, more current flows into the gate until the SCR is turned on. With the SCR on, the voltage across it falls almost to zero and the mains voltage appears across the lamp. The capacitor C is no longer being charged (since the SCR voltage is zero) and it just discharges into the gate of the SCR. At the end of the half cycle A, the mains voltage falls so that the current through the SCR falls below the holding current, causing the SCR to turn off.

During the negative half cycle B, the SCR does not allow current to flow, since current never flows through it in the reverse direction. The next positive half cycle C will repeat the same sequence of events as for A.

The lamp voltage is a series of part sine waves. Adjustment of VR_1 will alter the time taken to charge C, and hence the delay time shown in fig. 4.8 can be varied. The effect of varying the delay time on the lamp-voltage waveform is shown in fig. 4.9. Clearly, the shorter the delay time, the brighter the lamp will be, since the voltage across the lamp is present for a greater percentage of the time.

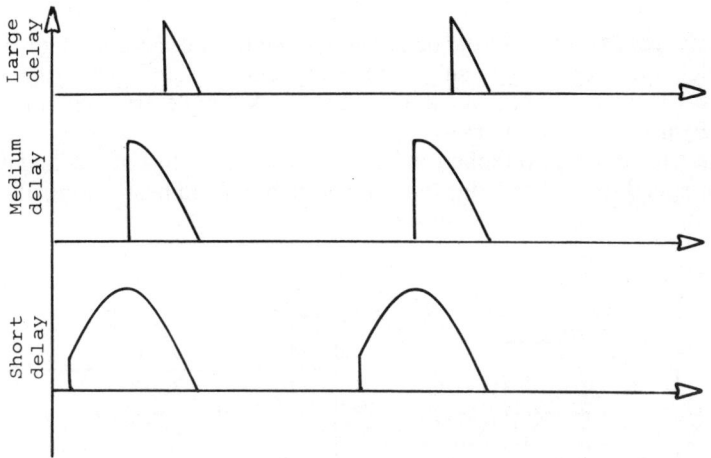

Fig. 4.9 Effect of delay time on lamp voltage

For the large delay time the voltage is present for only a very short period and the lamp will be virtually extinguished.

The term 'phase control' is used since we control the phase angle at which the SCR is fired. Despite the fact that the waveform supplied to the lamp is intermittent, it is switched at mains frequency which is too fast to be detected by the human eye. Similarly, in motor control this type of waveform presents no significant problems.

The triac

Where control of power in a.c. circuits is required, a single SCR cannot control power on positive and negative half cycles. This can be overcome by using two SCR's in inverse parallel connection as shown in fig. 4.10. SCR_1 conducts on the positive half cycle and SCR_2 on the negative half cycle.

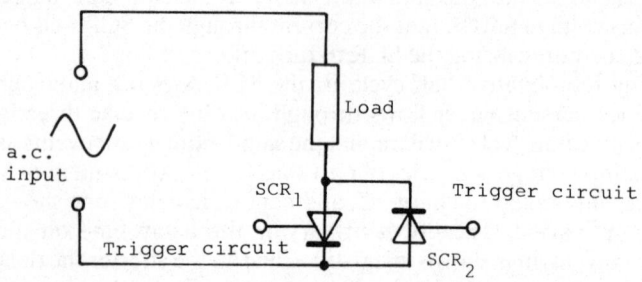

Fig. 4.10 Inverse parallel connection of SCR's

The triac, a three-terminal thyristor, is essentially two SCR's in inverse parallel connection with a single gate connection which can be triggered 'on' with a gate signal of either polarity. A simple triac circuit and waveforms are shown in fig. 4.11.

The triac is a very popular device for full-wave control of a.c. and d.c. motor speed, light dimming, static a.c. switching, and heater control.

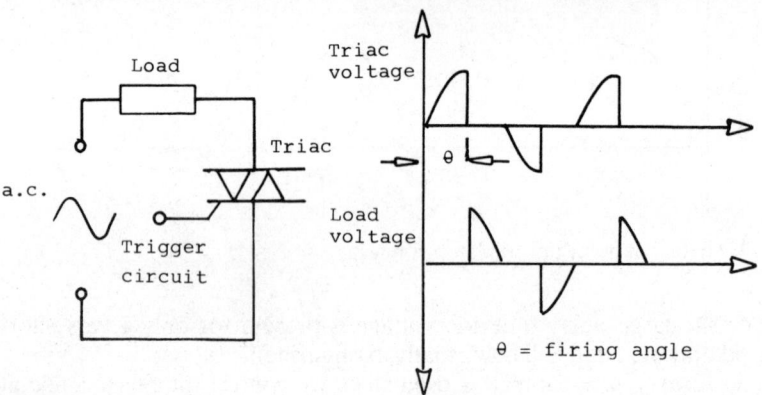

Fig. 4.11 Triac symbol, basic circuit, and waveforms

40

SCR and triac switching

Switching SCR's and triacs part way through a cycle, particularly when controlling large powers, can cause large transients (voltage spikes) on the power-supply lines. This noise can cause malfunctioning of other circuits, particularly digital circuits, on the same supply.

To overcome this problem, a method of triggering called 'zero-voltage switching' is gaining popularity. In this method, the SCR or triac is switched on only as the supply voltage changes from positive to negative. To achieve power control, the device will be off for complete cycles; i.e. control is effected by skipping whole cycles.

5   Optoelectronic devices

Introduction

Optoelectronic devices are simply specially made diodes and transistors which interact with light to a useful extent. All semiconductor devices react to light to some degree – one of the functions of their packages is to shut out light – but optoelectronic devices are designed to make efficient use of this phenomenon.

An optoelectronic device can be defined as one which

a) detects and/or is responsive to 'light', or
b) emits or modifies 'light', or
c) uses 'light' for its internal operation.

For these purposes, 'light' is taken to include energy in the ultra-violet and infra-red wavelengths as well as visible light.

There are two important categories of optoelectronic devices: light sensors and light emitters.

Some of the symbols currently in use are shown in fig. 5.1.

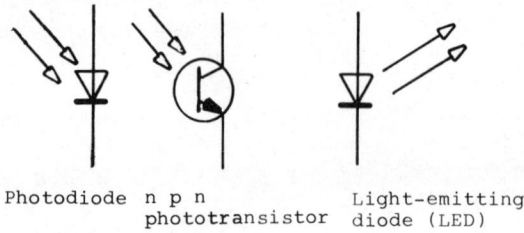

Photodiode   n p n   Light-emitting
              phototransistor   diode (LED)

Fig. 5.1 Some optoelectronic device symbols

Light sensors

A light sensor is simply a device which undergoes a reversible electrical change when exposed to light of the correct energy.

Photoconductors – photoresistors
When photodiodes are used *with* a separate power supply, they are called 'photoconductive', meaning that they conduct when illuminated and block current when dark.

Photovoltaic devices

When photodiodes are used to generate current *without* the assistance of another power supply, they are called 'photovoltaic', e.g. solar cells.

Phototransistors

This is a class of light-sensitive transistors in which the collector–base junction is illuminated to produce flows of electrons and holes which are then amplified in the normal way.

Other photosensitive devices include photoFET's, photo-Darlingtons, and photothyristors.

Light emitters

Light-emitting diodes simply reverse the effect used in light sensors.

Silicon and germanium are not suitable substances for light emitters: instead, a compound of elements is used as the semiconductor substance, the most popular types being gallium arsenide and gallium phosphide.

Applications of optoelectronic devices

Probably the most common use for optoelectronic devices is to read punched cards and tape.

Each hole in the punched card or tape represents one 'bit' of data. The card (or tape) is passed between an array of light emitters on one side and light sensors on the other. Each emitter generates light continuously, but each sensor is actuated only when a hole passes by and admits light to it. In this way, each hole transmits light to a sensor and creates a pulse of electricity. This type of system involves no moving parts.

This principle – interrupting a beam of light between an emitter and source – is the basis for many safety devices, e.g. to stop machinery if the operator breaks a suitably placed beam (usually infra-red). As the reliability and cost are improved, more and more industrial uses are being found, e.g. to count objects on conveyor belts; in inspection; in flow control, weight control, and level indicators; and in many more applications.

Displays

Light-emitting diodes (LED's)

Optoelectronic devices are finding increasing use in alphanumeric displays. In this application, seven LED's are arranged as shown in fig. 5.2(a) to form what is known as a 'seven-segment display'. By using a suitable 'driver' circuit, each segment can be operated separately, permitting any combination of segments to be operated at the same time to form any numeral or letter of the alphabet.

Certain applications such as digital multimeter displays use a 'half-digit'. This is normally used in the most significant position (the left-hand one in a display) and it is normally only required to indicate 1. Thus only the two right-hand segments of the display in fig. 5.2(a) are required for the half digit, so reducing the cost of the display and its associated drivers.

LED displays may also be in the form of a 5 × 7 dot matrix as shown in fig. 5.2(b).

(a) Seven-segment display

(b) 5 x 7 dot matrix

Fig. 5.2 Light-emitting displays

LED's can be obtained in 2.5mm to 25mm character heights; in single and multi-character displays; and in red, green, and yellow colours. The driver–decoders may be integral or separate.

However, LED's require appreciable currents, and under certain background-lighting conditions they are difficult to see.

Liquid-crystal displays (LCD's)

These devices are becoming more commonly used as display devices, due to their low cost, low power requirements, and good contrast. They are formed as seven-segment displays similar to LED's, but their principle of operation is completely different. Basically, a small amount of the liquid is sandwiched between two glass plates and an electric field is applied to the plates. This field causes disturbances in the liquid which cause the *external* light to be reflected, giving a 'light display'.

Since LCD's do not generate or emit light, they are not efficient in low background lighting. Also, they require a.c. drives.

Opto-isolators (or couplers)

Combining a light-emitting source and a photodetector in the same package forms the family of devices known as opto-isolators. A simple circuit arrangement is shown in fig. 5.3.

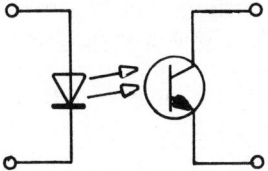

Fig. 5.3 Opto-isolator symbol

Opto-isolators are being used where electrical isolation between circuits is necessary and in digital applications where speed of operation and noise reduction are required.

6 Logic gates

What is a logic gate?

Any logic system, whether it is an industrial control or a computer, is made up of large numbers of simple units. These simple units are called 'gates', because one of their functions is to allow a signal to pass or not to pass, rather like opening or closing a gate.

What types of device can be used for a logic gate?

Many different technologies can be used to make a logic gate, such as

a) relays – although they are not often thought of in terms of gates;
b) electronics – valves, transistors, and integrated circuits can all be employed in logic gates;
c) pneumatics – systems using air can be made to perform logic. Such systems are generally referred to as 'fluidic'.

Since logical gating functions can be carried out by many different types of device, it is clear that a logic gate is an idea rather than a particular device. While our main interest will be in semiconductor electronic logic, it should be remembered that the same ideas and techniques apply to all logic.

Comparison with relay logic

Since many people are already familiar with relay logic, we shall use this to assist in explaining the general logic ideas. Firstly we must consider the differences.

Relay logic generally employs a large number of contacts on each relay coil. Electronic logic, on the other hand, works as if there was only one contact available on each coil. Thus interconnections between electronic logic units differ markedly from interconnections between relays. In electronic logic it is as if one contact were used to feed as many coils as required, i.e. one wire carrying a particular signal will feed all the logic units that require this information.

Logic symbols

The symbols used in this book are those used by most engineers and in most data books and are those of the United States MIL-STD-806B. Appendix 6 shows some other logic symbols.

Basic logic gates

The logic functions of the basic gates will be introduced in terms of relay logic with one contact per relay coil.

The AND gate

Clearly in order to energise RLO in fig. 6.1, both signals *A and B* must be at 5 V.

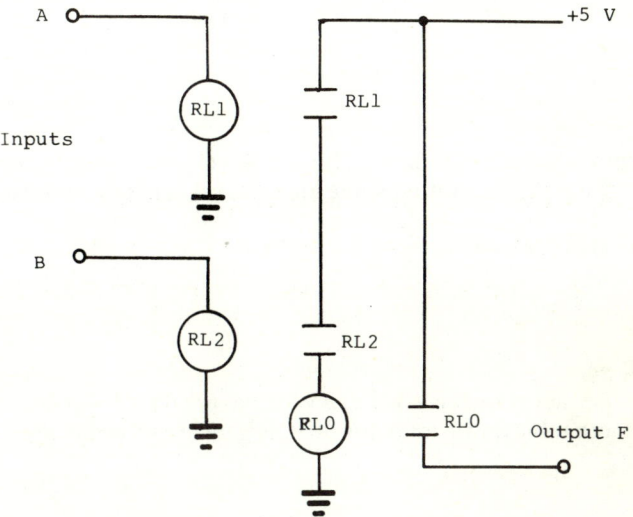

Fig. 6.1 A relay AND gate

Any circuit which requires all its inputs to be energised in order to produce an output is called an AND gate. Its symbol is given in fig. 6.2. (Note that *A*, *B*, and *F* do not form part of the symbol but are shown merely to illustrate how the symbol relates to the relay circuit.)

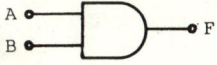

Fig. 6.2 The AND-gate symbol

It is usual to illustrate the function of a logic gate (or a combination of gates) using what is called a truth table. This table in fig. 6.3 gives the output condition for all combinations of inputs.

To make this type of table express the logic function more clearly, it is usual to replace the 0 V with just '0' and the 5 V with '1', as in fig. 6.4. This

47

A	B	F
0 V	0 V	0 V
0 V	5 V	0 V
5 V	0 V	0 V
5 V	5 V	5 V

A	B	F
0	0	0
0	1	0
1	0	0
1	1	1

Fig. 6.3 Voltage truth table for the AND gate

Fig. 6.4 Truth table for the AND gate

logic truth table can now be used for any AND gate – relay, electronic, or fluidic. (5 V is generally used as the logic-1 level as it is the standard supply for TTL devices – the 'industry standard' logic family, covered in more detail in chapter 14.)

The OR gate
Clearly if either *A or B* is a 1 (i.e. 5 V) in fig. 6.5 then *F* will be a 1.
    The truth table and symbol for the OR gate are shown in fig. 6.6.

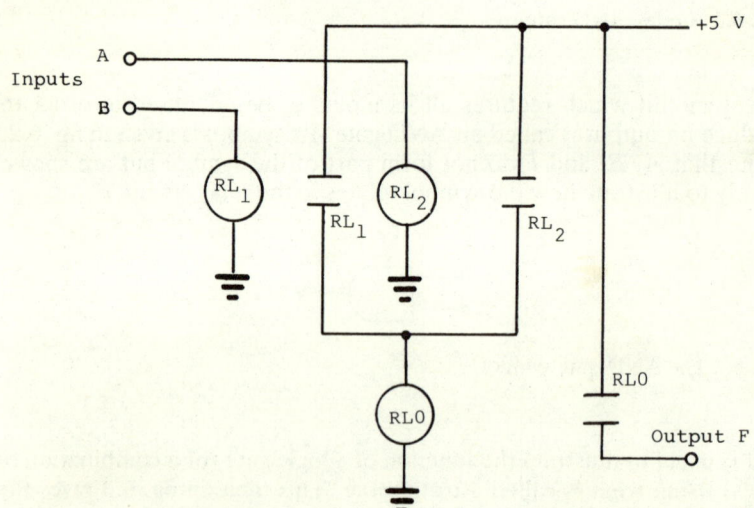

Fig. 6.5 A relay OR gate

A	B	F
0	0	0
0	1	1
1	0	1
1	1	1

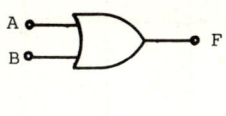

Fig. 6.6 Truth table and symbol for the OR gate

The NOT gate

This gate (fig. 6.7) is unusual in that it has only a single input. The output is always the opposite of the input and consequently the NOT gate is often referred to as an 'inverter'.

The truth table and symbol for the NOT gate are given in fig. 6.8.

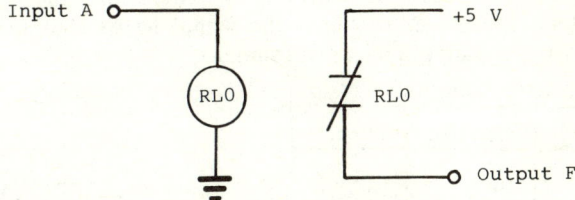

Fig. 6.7 The relay NOT gate

A	F
0	1
1	0

Fig. 6.8 Truth table and symbol for the NOT gate

Multiple-input gates

AND and OR gates (but not NOT gates) can have any number of inputs. They will still perform the same function; e.g. for the three-input AND gate shown in fig. 6.9, all inputs must be 1 to give an output of 1.

A	B	C	F
0	0	0	0
0	0	1	0
0	1	0	0
0	1	1	0
1	0	0	0
1	0	1	0
1	1	0	0
1	1	1	1

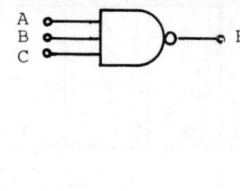

Fig. 6.9 Truth table and symbol for a three-input AND gate

Properties of gates

Gating

When the control signal in fig. 6.10 is set to 0, the output is always 0; but when the control is set to 1 the output signal is identical to the input signal. The effect is, therefore, to allow the signal to pass or not to pass depending on the state of the control signal.

Signal	Control	Output
0	0	0
1	0	0
0	1	0
1	1	1

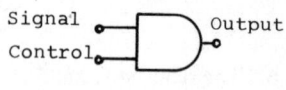

Fig. 6.10 The function of 'gating'

Similarly, an OR gate will perform the same sort of function but in this case the signal will not pass when the control signal is 1, when the output will also be 1.

Selection

Referring back to the truth table for the three-input AND gate (fig. 6.9), it can be seen that there is only one combination of the inputs which will give a 1 on the output. This combination is when all inputs are 1.

It is often required to be able to pick out a particular combination of signals which is not necessarily all 1's. This can be achieved by using NOT gates combined with AND gates.

The logic circuit in fig. 6.11 will produce a 1 on the output only when $A = 0$, $C = 1$, and $D = 1$. Any other desired combination could be picked out by using NOT gates on the appropriate inputs to the gate.

| | AND-gate inputs | | | |
A	B	C	D	F
1	0	0	0	0
1	0	0	1	0
1	0	1	0	0
0	1	0	0	0
1	0	1	1	0
0	1	1	0	0
0	1	0	1	0
0	1	1	1	1

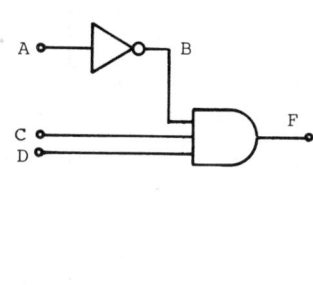

Circuit inputs

Fig. 6.11 The function of 'selection'

Note that the OR gate can also pick out particular combinations, since a 0 output is given only when all the gate inputs are 0.

Combination of gates

Gates can be combined so as to perform more complex functions than the simple AND, OR, and NOT. We have already seen a simple combination used for gates performing a selection function. The method used in this example represented the output of the NOT gate, and hence one input to the AND gate, by B. This method is generally useful.

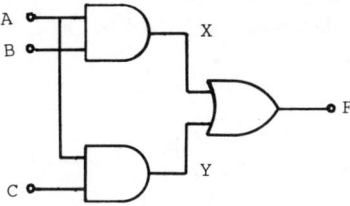

Fig. 6.12 Simple combination of logic gates

Consider the logic diagram of fig. 6.12. To find out how it operates, we shall use the truth-table method. Construct a truth table showing all combinations of all inputs A, B, and C. Name all intermediate signals, i.e. X and Y, and add columns for these signals to the truth table. The next column in the truth table is for the output F. The table is now completed: all combinations of A, B, and C are entered, from which we derive X and Y according to the rules for an AND gate, and finally we can arrive at F by using the OR-gate rules on X and Y. The completed table is shown in fig. 6.13.

51

Inputs			A AND B	A AND C	X OR Y
A	B	C	X	Y	F
0	0	0	0	0	0
0	0	1	0	0	0
0	1	0	0	0	0
0	1	1	0	0	0
1	0	0	0	0	0
1	0	1	0	1	1
1	1	0	1	0	1
1	1	1	1	1	1

Fig. 6.13 Truth table for the circuit of fig. 6.12

We can now restate the function of this circuit in words: the output F will be 1 when $A = 1$ and $B = 1$, or when $A = 1$ and $C = 1$.

An alternative approach to combinational logic is given in chapter 7.

NAND/NOR – gates for all purposes

The NAND gate

A particular combination of two gates is called a NAND gate. As its name suggests, it is a combination of an AND gate with a NOT gate. The function and symbol are given in fig. 6.14.

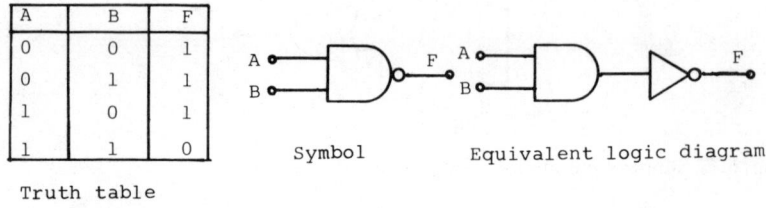

A	B	F
0	0	1
0	1	1
1	0	1
1	1	0

Truth table

Symbol   Equivalent logic diagram

Fig. 6.14 The NAND gate

The NAND is important since it is the usual logic function to be performed by an integrated-circuit gate. The NAND gate is also remarkable in that it can be used to perform the functions of AND, OR, and NOT gates. This is illustrated in fig. 6.15, where the inputs shown as 1 are connected to a permanent logic 1.

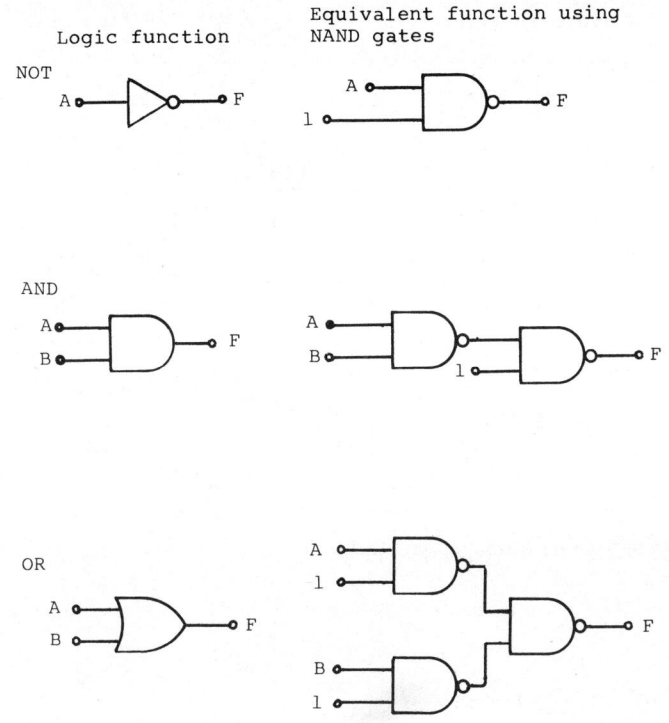

Fig. 6.15 NAND – the 'all-purpose' gate

The NOR gate

A similar gate to the NAND is the NOR – a combination of an OR with a NOT gate. The function and symbol for the NOR gate are given in fig. 6.16, and its use for AND, OR, and NOT functions is shown in fig. 6.17, where the inputs shown as 0 are connected to a permanent logic 0.

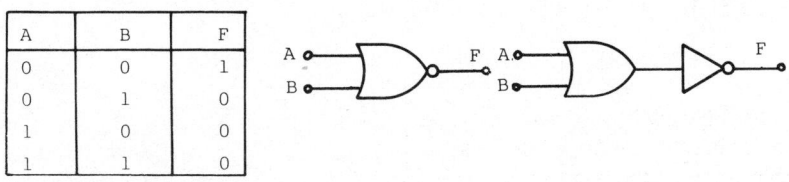

A	B	F
0	0	1
0	1	0
1	0	0
1	1	0

Fig. 6.16 The NOR gate

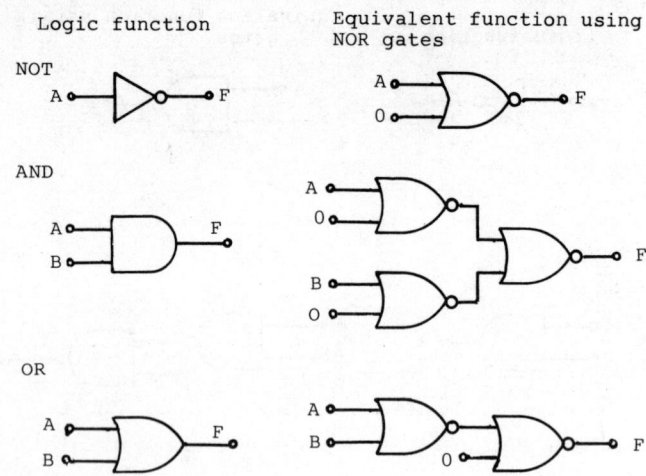

Fig. 6.17 NOR – the 'all-purpose' gate

7   Boolean algebra

Introduction

In 1847 a man called George Boole wrote a pamphlet with the title *The mathematical analysis of logic – being an essay toward a calculus of deductive reasoning*. This pamphlet gathered dust until in 1938 a man called Claude Shannon realised that the form of mathematics devised by Boole could be applied to relay switching circuits. Boolean algebra has now become a most important method of dealing with electronic logic.

The algebra

The OR function

The signal produced at the output of an OR gate is shown by the use of a + sign between the inputs. Note that the + sign does not denote normal addition but represents the OR function. Some people use 'v' instead of '+' to avoid this problem.

The output of an OR gate whose inputs are *A* and *B* is ($A + B$). This is shown in fig. 7.1.

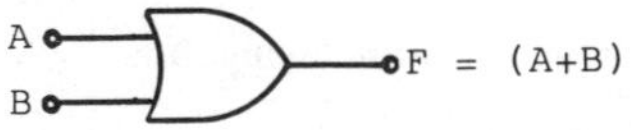

Fig. 7.1 The OR gate

The AND function

The AND function is indicated by a dot. Note that this dot indicates the AND function and not normal multiplication. The use of the dot is shown in fig. 7.2.

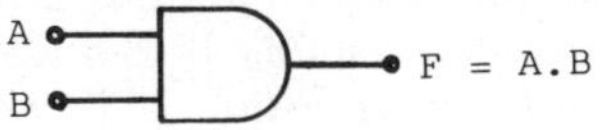

Fig. 7.2 The AND gate

The NOT function

The NOT function is indicated by a bar over the signal name, as in fig. 7.3.

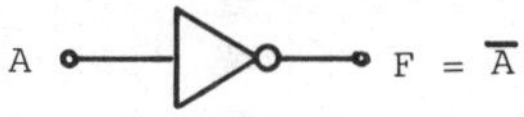

Fig. 7.3 The NOT gate

The NAND and NOR functions

These functions are built up by combining the previous Boolean symbols as shown in fig. 7.4.

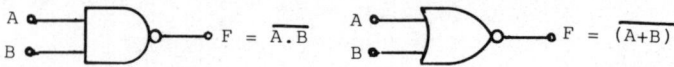

Fig. 7.4 NAND and NOR gates

It should be carefully noted how the bar is used in this case. The bar appears over the whole signal name and not over the individual signal names. Writing $\bar{A}.\bar{B}$ is entirely different from $\overline{A.B}$.

More complex signals

More complex signals can be built up by starting with the signal inputs and writing down the output of each gate in turn until the final output is reached. An example is shown in fig. 7.5.

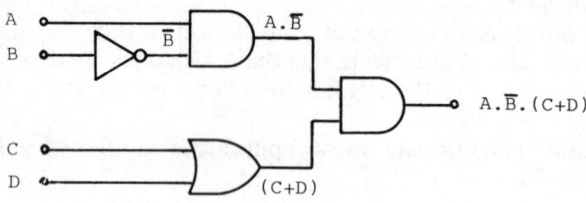

Fig. 7.5 An example of the use of Boolean algebra

This output signal can be rewritten in a different form using normal arithmetic rules for multiplying brackets and treating the dot and plus signs *as if* they were multiplication and addition,

i.e. $(A.\bar{B}).(C+D) = A.\bar{B}.(C+D) = A.\bar{B}.C + A.\bar{B}.D$

The value of this algebra is that we can find out whether the output signal will be 0 or 1 for any combination of 1's and 0's on the inputs, without referring to the logic circuit itself. Boolean expressions are often shown on logic diagrams for this reason.

Basic rules of Boolean algebra

To find the output state from the algebra, it is necessary to learn some of the rules of Boolean algebra. They are

1. $X+0 = X$ 4. $X.1 = X$ 7. $X.X = X$

2. $X+1 = 1$ 5. $\bar{\bar{X}} = X$ 8. $X+\bar{X} = 1$

3. $X.0 = 0$ 6. $X+X = X$ 9. $X.\bar{X} = 0$

If we consider the first of these, then any signal X which is ORed with a logic 0 will give the same signal X on the output of the OR gate. The same sort of argument applies to each of the rules up to the last. The fifth expression merely states that the result of putting a signal X through two inverters is to reproduce the original signal.

To see how this works let us consider the previous example of an output signal $(A.\bar{B}.C + A.\bar{B}.D)$.

Example 1   If input A is 0, the first term $A.\bar{B}.C = 0.\bar{B}.C = 0.(\bar{B}.C)$. If we consider $(\bar{B}.C)$ as the signal X in rule 3, then $(A.\bar{B}.C) = 0$. Similarly $A.\bar{B}.D = 0$. Thus $A.\bar{B}.C + A.\bar{B}.D = 0$.

Example 2   If $A = 1$ and $\bar{B} = 0$ and $D = 1$ then $A.\bar{B}.D = 1$ and so $A.\bar{B}.C + A.\bar{B}.D = A.\bar{B}.C + 1 = 1$, where $A.\bar{B}.C$ is the signal X in rule 2.

Note that it is not always necessary to consider the state of all inputs to determine the state of the output. It is this fact which makes the Boolean algebra valuable.

The above (with sufficient practice) should be adequate to meet most test and repair requirements but, if required, further information is available in any good logic textbook.

8 Sequential circuits

Introduction

The logic devices discussed previously are known as 'combinational circuits' and their outputs depend solely on the inputs – not on the sequence in which the inputs are applied, nor on the state of the circuits before the inputs are applied. Circuits whose outputs *do* depend on the sequence in which the inputs are applied or on the previous state of the circuit are known as 'sequential circuits' and fall into two types: *synchronous*, where the change in the output is synchronised with a clock, and *asynchronous* where the output is independent of the clock.

To keep track of a sequence of events, a device having the capability to remember things (a memory) is required. There are many forms of memory or storage, such as magnetic core stores and magnetic tapes used in computers. The most usual memory device in industrial control systems is an electronic circuit called a 'flip-flop'. There are various types of flip-flop, and some of the most important are covered below.

The set–reset flip-flop

This basic form of flip-flop, known as an S–R flip-flop, can be readily built up using NAND gates (which further demonstrates the versatility of the NAND gate) as in fig. 8.1, which also shows a NOR version.

When a 1 is present on the S (set) input, the output Q will become 1 ($\overline{Q}$ becomes 0). When the 1 on the S returns to 0, Q will remain at 1, i.e. the circuit 'remembers' that S was previously 1. When R (reset) is a 1 then the output Q is reset to 0 ($\overline{Q}$ becomes 1), i.e. the circuit is told to 'forget'.

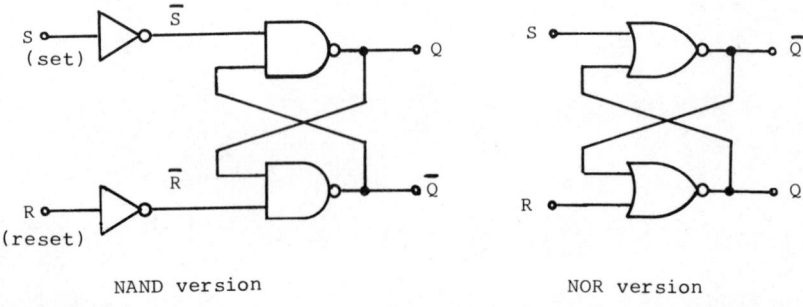

Fig. 8.1 The S–R flip-flop

Often this type of flip-flop will be found without the inverters on the input, so that the circuit is 'set' by a 0 on the S input and 'reset' by a 0 on the R input.

The logic level which causes a device to operate or change state is often referred to as the 'active' input level – this may be either 0 or 1, depending on the device construction. *Both* active levels may be present on different inputs on the same device, thus the use of the term 'active' avoids confusion. A 'bubble' or circle is used on logic symbols to indicate that the input is active 'low'.

There is a peculiar condition on the S–R flip-flop. Suppose we attempt to both set and reset the flip-flop at the same time. Both outputs will be a 1 and, as the outputs are no longer the inverse of each other, the names Q and Q̄ are inappropriate (despite this, they are in common use). When the set and reset signals are removed together, there is no way of telling what the output will be.

S	R	Memory output Q
0	0	No change
0	1	0
1	0	1
1	1	Indeterminate

Fig. 8.2 Truth table for the S–R flip-flop

All the functions of the flip-flop can be shown in a truth table, fig. 8.2. The output column is headed 'memory output', to indicate the state of the output remaining after S and R have both returned to 0.

Except for the indeterminate case, Q̄ is always the reverse of Q.

D-type flip-flops

A more widely used flip-flop is the D-type flip-flop, the general symbol for which is shown in fig. 8.3.

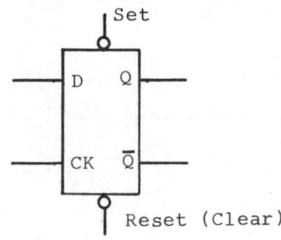

Fig. 8.3 D-type flip-flop symbol

D	Clock	Q	$\overline{Q}$
0	⎍	0	1
1	⎍	1	0

Fig. 8.4 Truth table for the D-type flip-flop

If only the set and reset inputs are used, the device functions as a simple S–R flip-flop. If the D and clock (CK) inputs are used, it can be seen from the truth table in fig. 8.4 that the input on the D input is transferred to the Q output after a clock pulse.

The J–K flip-flop

The J–K flip-flop does not have the indeterminate output state of the S–R flip-flop and the two outputs are always opposite, so that Q and $\overline{Q}$ are the correct designations.

While a J–K flip-flop can be built up using NAND gates, the circuit is complex and does not aid explanation. The J–K is generally drawn using its own symbol as in fig. 8.5.

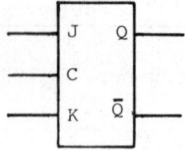

Fig. 8.5 J–K flip-flop symbol

This flip-flop has three inputs: J, K, and C (sometimes the C input is called T). The J and K inputs are similar to the S and R inputs on the S–R flip-flop, but the outputs of the J–K flip-flop cannot change until a pulse appears on the C input. A timing diagram in fig. 8.6, similar to the display on an oscilloscope, illustrates the type of pulse required on C and indicates the part of the pulse on which the Q output would change.

The way in which the output changes depend on the J and K inputs is given in the truth table in fig. 8.7.

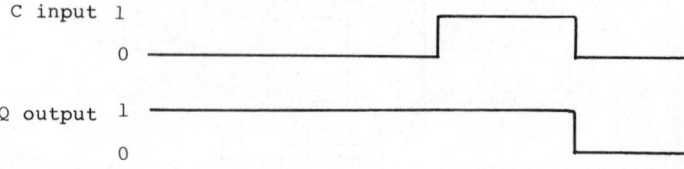

Fig. 8.6 Timing diagram

60

J	K	Output after one clock pulse	
0	0	No effect	
0	1	Q = 0	$\overline{Q}$ = 1
1	0	Q = 1	$\overline{Q}$ = 0
1	1	Changes Q to opposite state	

Fig. 8.7 Truth table for the J–K flip-flop

In practice, most commercial J–K flip-flops are also provided with S and R inputs. These additional inputs operate on the outputs just like the previously described S–R flip-flop. This increases the versatility of the device and an integrated-circuit form is easily provided. The symbol is given in fig. 8.8.

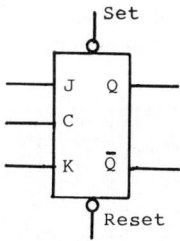

Fig. 8.8 J–K flip-flop with 'set' and 'reset' ('clear') inputs

In use the S and R inputs will override the other inputs – i.e. if R = 1 then Q will be 0 no matter what inputs are present on J, K, and C; but normal J–K operation takes place when R = 1 and S = 1.

J–K master–slave flip-flops

In practice, the simple J–K flip-flop, fig. 8.9, suffers from the disadvantage that if the clock is still high and the J or K inputs change, the flip-flop will be immediately set or reset. This problem is overcome by using a master–slave J–K flip-flop. For normal operation the set and clear must be inactive, i.e. high or logic 1; the device will then change state on the negative or trailing edge of the clock.

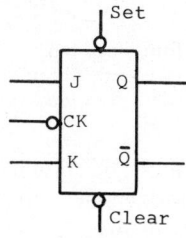

Fig. 8.9 J–K master–slave flip-flop

9 Counters and shift registers

Binary counters

The logic diagram for a four-stage binary counter is given in fig. 9.1. The term 'four-stage' is used to signify the number of flip-flops used. The Q output of each flip-flop is used to provide the C input pulse to the next flip-flop. The J and K inputs are shown disconnected, which in most logic modules gives the same effect as connecting them to a 1. Whenever the C input of one of these flip-flops moves from 1 to 0, the flip-flop output will change to its inverse. This is shown in the timing diagram in fig. 9.2.

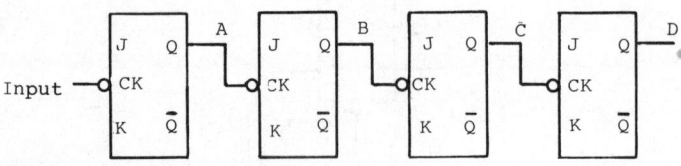

Fig. 9.1 A four-stage binary counter

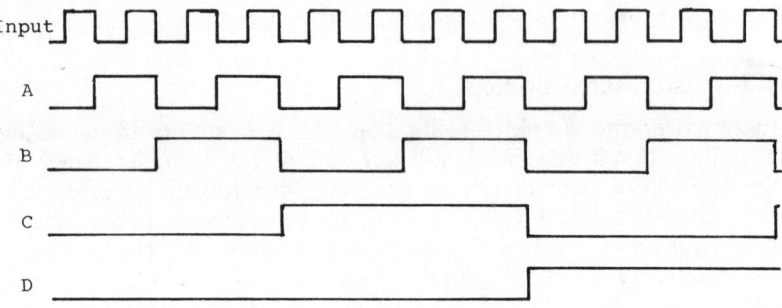

Fig. 9.2 Timing diagram for the four-stage binary counter

This circuit is called a 'counter', because it counts the input pulses. The counting is not in the familiar decimal system but in binary. The decimal system is based on 10 and uses the ten characters between 0 and 9 for each column of units, tens, hundreds, etc. In logic we only have the two numbers 0 and 1, so the counting is based on 2. The method of counting is shown in fig. 9.3.

Binary	Decimal
0	0
1	1
10	2
11	3
100	4
101	5
110	6
111	7
1000	8
1001	9
1010	10
etc.	etc.

Fig. 9.3 Comparison of binary and decimal counting

Note that if we consider the right-hand column of the binary count as 1, the next column as 2, the next as 4, and the next as 8 (i.e. doubling each time), we can readily translate a binary number into decimal by adding up the appropriate numbers each time as 1 appears, e.g. 10110 in binary $= 16 + 4 + 2$ in decimal $= 22$ in decimal.

Shift registers

A shift register (often referred to simply as a 'register') provides a store for a binary number together with the ability to move the data from left to right each time a shift pulse is applied. A five-stage shift register is illustrated in fig. 9.4. The timing diagram for this shift register is shown in fig. 9.5 for the case when the register contains the number 11001, the input is kept at 0, and five shift pulses are applied.

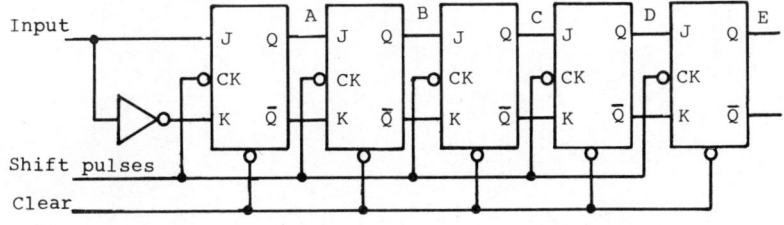

Fig. 9.4 A five-stage shift register

63

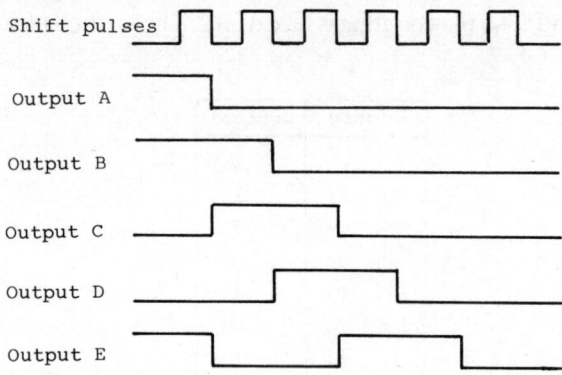

Fig. 9.5 Timing diagram for the five-stage shift register

The timing diagram shows how the original number is moved one place to the right for each shift pulse. One digit is lost for each shift pulse until after five pulses the counter contains only 0's.

The way each flip-flop changes is defined by the previous flip-flop, whose Q and Q̄ outputs are connected directly to the J and K inputs. Data is thus transferred from one flip-flop to the next when the shift pulse appears.

Data can be entered into the shift register by using the input in combination with shift pulses and moving the data down the register until it reaches the desired position. Data can also be entered directly into each flip-flop individually by using the S and R inputs which are often provided in addition to the J and K inputs. These S and R inputs operate exactly like the inputs described on the S–R flip-flop, i.e. they can cause the flip-flop to change at any time without a clock pulse.

10 Control circuits

Introduction

Sequential circuits are very commonly used in industrial control. They are circuits in which time plays a part and the controlled functions happen in a time sequence. These circuits clearly involve memory of past events which are stored in flip-flops (sometimes called bistables) but, in addition, circuits are required which provide an output for a specified time, e.g. to control spot-welding time. One circuit which can perform this function is called the 'monostable'.

The monostable

Functions of monostables

A monostable is initiated by the edge of a pulse and once initiated produces an output pulse of a pre-set width. This is illustrated in fig. 10.1.

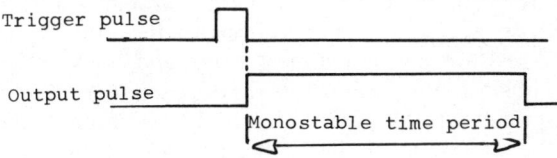

Fig. 10.1 Monostable timing diagram

Monostables can be triggered (initiated) by the negative-going edge of the trigger pulse as shown in fig. 10.1 or by the positive-going edge of the trigger pulse, depending on the design of the monostable circuit. A monostable triggered on positive-going edges is illustrated in fig. 10.2.

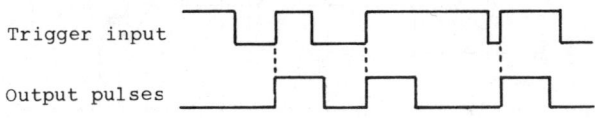

Fig. 10.2 Monostable timing diagram with a positive-going trigger

Monostables are often 'cascaded', so that when one monostable time period ends it triggers a second monostable. A number of monostables can be cascaded in this way to perform a sequence of operations. An example is given in fig. 10.3 for a spot welder. The 'initiate the weld sequence' signal triggers the 'clamp' monostable which allows sufficient time for the clamps to operate. The end of the clamp monostable's time period triggers the 'weld' monostable, and the welding current flows while the output from this monostable is present. At the end of this time, the 'unclamp' monostable is initiated by the negative-going edge of the 'weld' monostable and the clamps are given time to release.

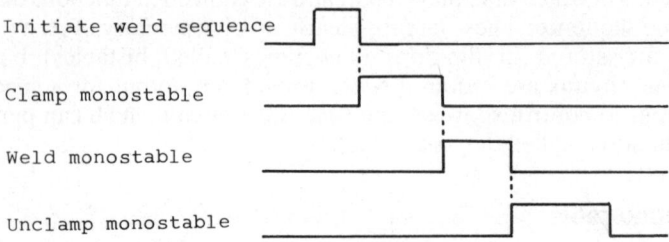

Fig. 10.3 Monostable used to control spot-welding sequence

Practical monostables
Virtually all monostables use the discharge through a resistor of the charge stored on a capacitor as the basis of the time period. Large capacitors and resistors give long time periods, and vice versa.

A circuit using discrete components (as opposed to a circuit produced in integrated-circuit form) is given in fig. 10.4. It is similar to the flip-flop circuit with the exception of the timing components R_T and C_T. The circuit normally sits with T_1 turned on and T_2 turned off. Note that in this state C_T will be charged up, since one side is connected to the

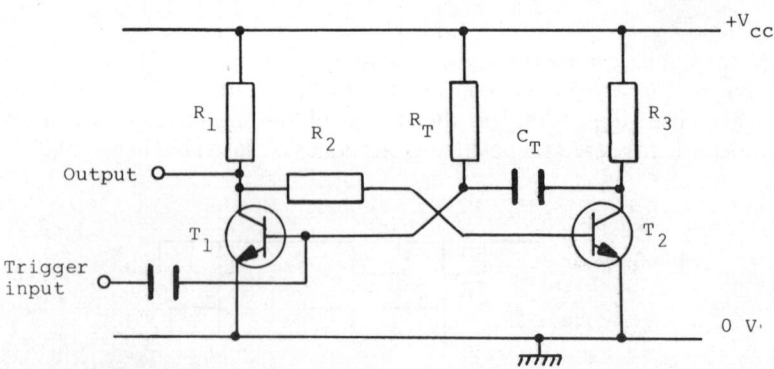

Fig. 10.4 Monostable circuit

66

supply voltage through R_3 and the other side is connected to the base of T_1 at 0.6V.

A negative-going signal on the trigger input will cause a negative-going signal on the base of T_1 so that T_1 will be turned off. Due to normal bistable action, if T_1 is off T_2 will be turned on. As T_2 turns on, its collector voltage falls from V_{cc} to almost 0V.

Since voltages across capacitors cannot be changed instantly, the base voltage on T_1 will be taken negative by the capacitor C_T.

T_1 will now remain off because of this negative voltage on its base. The capacitor C_T now starts to discharge through R_T, so the negative base voltage on T_1 gradually becomes more positive. Eventually, when the base voltage has risen far enough, T_1 turns on again, the circuit resumes its normal state, and the capacitor C_T charges up again through R_3. Waveforms for this circuit are given in fig. 10.5.

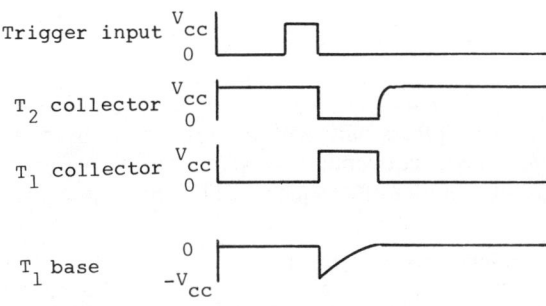

Fig. 10.5 Waveforms for fig. 10.4

Note that the voltage on the base of T_1 goes negative by a voltage almost equal to V_{cc}. Most modern transistors have a maximum reverse V_{be} of 5V, so if V_{cc} is greater than 5V the transistor is overrated and may eventually fail.

One way of constructing a monostable using NOR logic is shown in fig. 10.6. The NOR gates A and B form a bistable. A logic 1 applied to the

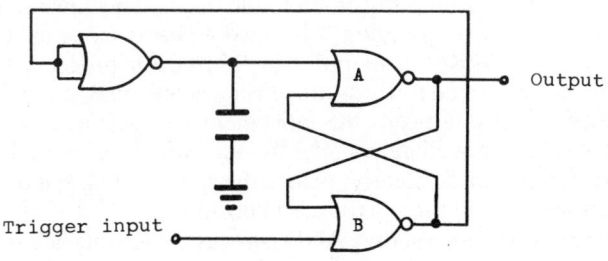

Fig. 10.6 Monostable using NOR logic

trigger input forces the output of B to 0. The output of the inverter, which was logic 0, attempts to change to 1 but is held back by the capacitor. Only when the capacitor charges up does the input to A become a 1, changing the bistable back to its original state. The timing diagram in fig. 10.7 illustrates the operation of this monostable.

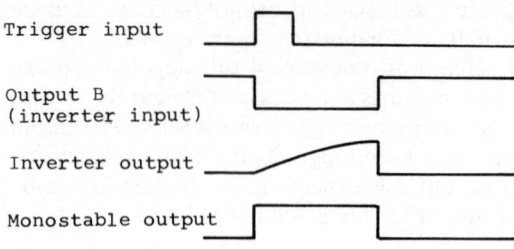

Fig. 10.7 Timing diagram for fig. 10.6

In practice, monostables built with discrete components are not very common, since more convenient versions can be constructed using integrated-circuit modules units such as TTL (see chapter 14).

Another type of sequential circuit

Sequential circuits involving monostables used as in the spot-welding example previously have serious drawbacks. The main problem is that, if the clamps do not operate in the time allowed for this to happen, the sequence will continue with possible disastrous consequences.

There is another type of sequential circuit which checks that each operation is completed before proceeding with the next one. This is inherently safe because, if part of the controlled machine fails to operate, the sequence will stop until it is corrected. Many industrial controls use a mixture of both systems so that an alarm is signalled if an operation is not completed within a pre-set time.

An example of the second type of system is shown in fig. 10.8. We shall consider the logic required to complete the operation shown in fig. 10.8 and to initiate the next operation, which would be to spot weld the three sections together. While the operation in fig. 10.8 is in progress, the three sections are being moved towards the sensing heads A, B, and C. When a part reaches its sensing head, the movement is automatically stopped. Since the parts are moved independently, they will arrive at their sensing heads at different times. Clearly, before the spot welding can be started, we must ensure that all three parts are in position.

This can be done with a simple AND gate, as the sensing heads give out a logic 1 when they detect the presence of the section. This is illustrated in fig. 10.9. The signals from the sensing heads are called A, B, and C, so the

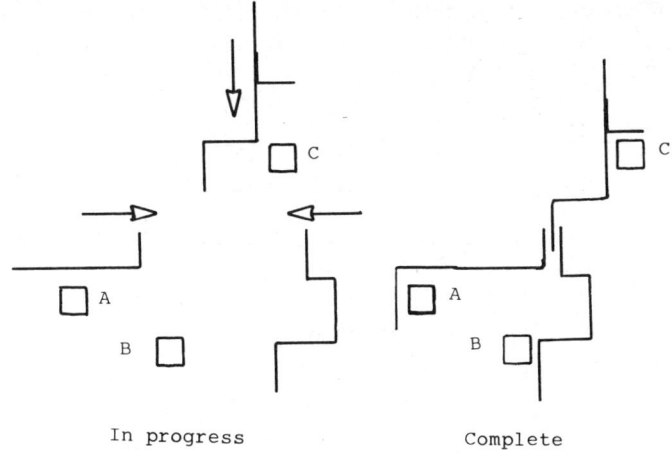

In progress   Complete

Fig. 10.8 Operation cycle

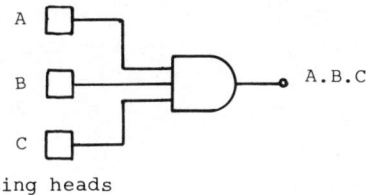

Sensing heads

Fig. 10.9 Detecting all parts in position

AND-gate output will be $A.B.C$, which is a logic 1 only when A *and* B *and* C are each logic 1. This is the condition to allow the spot-welding cycle to begin, so the signal $A.B.C$ can be used to initiate the spot welding when it becomes a logic 1.

For requirements later on in the system, it is necessary to remember that this operation has been completed. The signal $A.B.C$ is therefore fed into a flip-flop, and it is the output of the flip-flop which initiates the welding – see fig. 10.10.

Once we have a flip-flop remembering that the operation is complete, it will continue to do so for ever unless it is reset – hence the addition of a '+ reset' signal.

The use of + and − signs before the signal names is to clarify signal functions. The sign indicates the voltage that would be measured on that line when the signal is performing the operation stated in the signal name. For example, a '+ reset' signal will be performing the reset function when the voltage is most positive (+5V in TTL). Similarly, a signal '− start' would perform the start function when the voltage is most negative (0V in TTL).

69

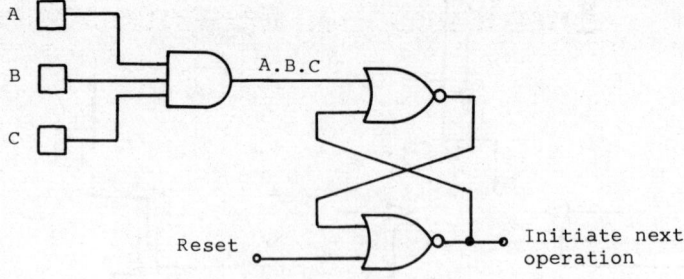

Fig. 10.10 All parts in position, remembered

11   Introduction to integrated circuits

What is an integrated circuit?

Integrated circuits (IC's) are complete electronic circuits containing resistors, capacitors, diodes, transistors, and their associated electrical interconnections manufactured on a single chip of silicon. Figure 11.1 shows typical packages used for IC's.

Fig. 11.1 IC packages (coin diameter 28 mm)

The active part of the device is a tiny chip of silicon, which may be no bigger than that for a single discrete transistor such as a BC 109. Devices such as microprocessors do come in larger sizes, but most of the increased size is necessary to enable connections to be made to the outside world. The connecting leads must be rigid enough to permit them to be inserted into IC holders. The complete package is extremely rugged.

A typical chip is shown greatly magnified in fig. 11.2. This is a very simple chip – most IC chips are bigger. Some of the components are identified in fig. 11.2: T = transistor, R = resistor, and D = diode.

71

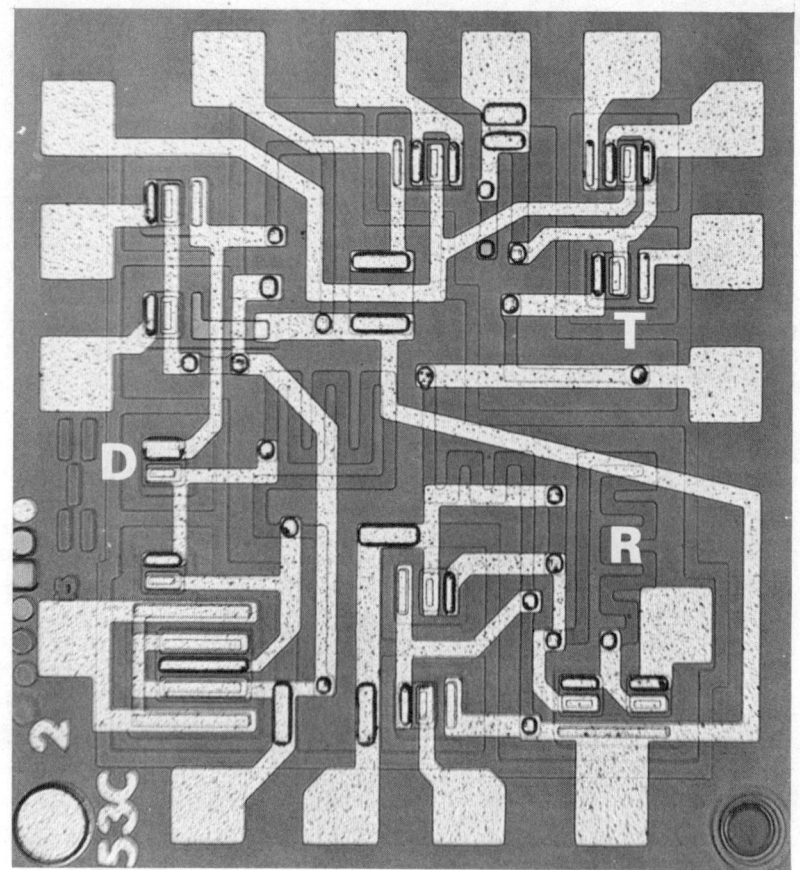

Fig. 11.2 IC chip

The electrical circuitry within the IC could be constructed with discrete components, but the advantages of IC's over conventional discrete circuits – e.g. small size, cheapness, reliability, low power consumption etc. – enable the device to be considered simply as a building block whose operation is modified according to external components which are added to achieve the required effect on the input signal.

Integrated-circuit limitations

The major limitation of the IC is its inability to dissipate large amounts of power, due to its small size. Any increase in current produces heat which may destroy the device, by melting the active regions. Thus IC's are usually limited to information processing, and the 'working' or output stages are usually discrete components.

Due to the small physical size of the device, the separation between conductors is very small, restricting the voltages which may be applied.

Manufacturing techniques do not permit accurate control over component values, and at present inductors and transformers cannot be produced on IC's.

Manufacture of integrated circuits

IC's are produced hundreds at a time on a slice of pure silicon about 5cm in diameter. Once the process is complete, the wafer is cut into individual circuits which are then mounted and encapsulated.

The individual components and circuits are built up in layers, producing a three-dimensional circuit.

The 'artwork' for each layer is produced using an accurately made 'photo mask' and is then reduced by photographic means for the entire slice with hundreds of identical circuits on it.

After each process step, the slice is left with a layer of oxide over its surface, and 'windows' are etched into this layer to permit the subsequent introduction of impurities into the slice. The sequence of operations is as

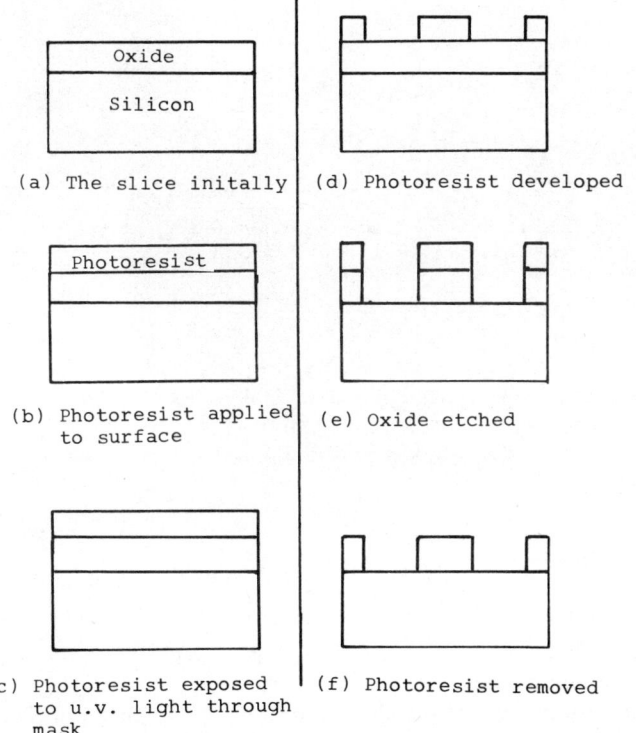

(a) The slice initally   (d) Photoresist developed

(b) Photoresist applied   (e) Oxide etched
    to surface

(c) Photoresist exposed   (f) Photoresist removed
    to u.v. light through
    mask

Fig. 11.3 IC etching sequence

73

follows. A compound called 'photoresist' is applied to the slice to form a thin even coating. The mask, consisting of a pattern of opaque areas on a glass sheet, is placed on the slice and u.v. light is shone through it. This has the effect of hardening the exposed areas of photoresist. The soft unexposed areas can then be removed by a solvent. Then, by etching with an acid, the exposed oxide areas can be selectively dissolved. Following this step, the remaining photoresist is also removed. The pattern of the mask has then been transferred to the slice.

Once the windows have been formed in the oxide, the slice is placed in a diffusion furnace and heated to about 1200°C in an atmosphere containing a suitable donor impurity. The donor penetration depth is determined by the length of time the slice is exposed in the furnace.

These intricate steps are repeated for each successive layer required until the whole circuit has been fabricated.

The window-etching sequence is shown diagrammatically in fig. 11.3, and fig. 11.4 shows a typical slice after processing.

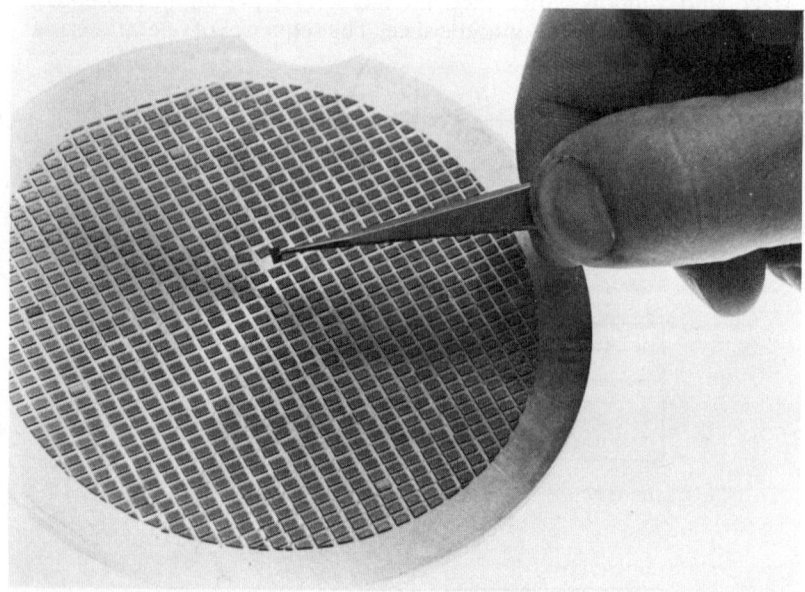

Fig. 11.4 Processed IC slice

Mounting and bonding
Once the processing of the slice is complete, it is cut up into individual circuits and the electrical connections are made by vibration welding. The IC is then encapsulated in a suitable package to provide protection.

Testing integrated circuits

After assembly, each IC is put through a series of tests of its electrical performance to make sure it meets the data-sheet specifications over its working temperature range.

IC's are usually tested first in the range 0 to 70°C, which is satisfactory for most commercial and industrial applications. IC's which pass this test are then usually tested in the range $-55°C$ to $+125°C$, and circuits passing tests in this range are then suitable for most military or space applications. This provides two grades of IC (refer to chapter 14 for further explanation).

12   Digital integrated circuits

Purpose of digital integrated circuits

Digital IC's are devices whose basic function is to manipulate binary or digital data by means of switching circuits.

Any electronic operation is capable of being carried out by digital circuitry. Although the digital circuits are generally more complex than their linear (analogue) equivalents, mass production of cheap reliable digital circuits ensures that this is not a problem.

Digital circuits have the major advantage that they can tolerate a large amount of distortion before they become unrecognisable, and they can easily be reconstructed to their original shape. The use of error-detecting and correcting codes enables errors due to noise to be removed. When used with binary digits, these codes permit the transmitted data to be checked and, if an error is found, the incorrect bit to be pin-pointed. This is achieved using a combined horizontal and vertical coding system.

Any digital electronic system to process or store information – such as computers, calculators, watches, process and machine-tool controls – is made up from a number of standard building blocks which are available in commercial packages. These are circuits which carry out specific functions, e.g. basic gates (NAND etc.), J–K flip-flops, adders, counters, and shift registers.

Digital systems usually consist of two distinct parts: a memory and a decision-making part.

Typical digital building blocks

Some commonly available devices are described briefly below.

Monostables
The monostable takes an input and produces an output delayed by a time t_d which is determined by external timing components. This is shown in fig. 12.1.

Schmitt trigger
The Schmitt trigger is a basic logic gate which can be used as a simple 'clock', fig. 12.2(a), or to reconstruct a distorted pulse, fig. 12.2(b). This device operates on a voltage level on the input rather than on a rising or falling edge.

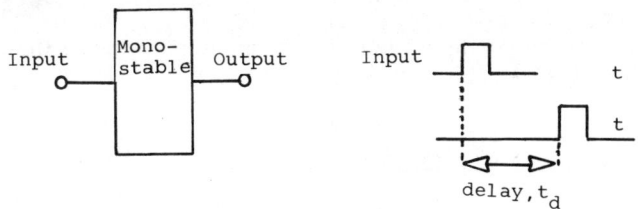

Fig. 12.1 Monostable

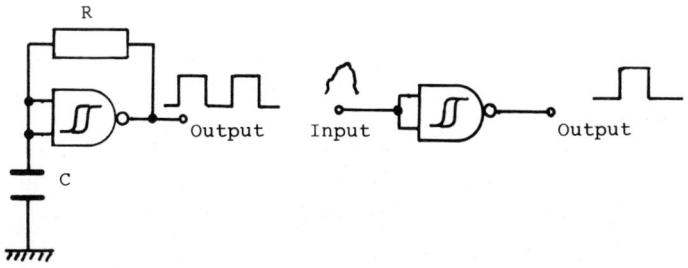

(a) Schmitt-trigger clock   (b) Level restore (pulse reshape)

Fig. 12.2 Schmitt-trigger circuits

Buffers

Buffers are very important devices for interfacing, or connecting different systems together.

Of particular importance are three-state buffers, or latches. These devices permit many outputs to be connected in parallel. They have three possible output states:

a) high (logic 1);
b) low (logic 0);
c) output disabled – connected neither to supply nor to ground, i.e. open circuit and hence high impedance.

This is shown in fig. 12.3.

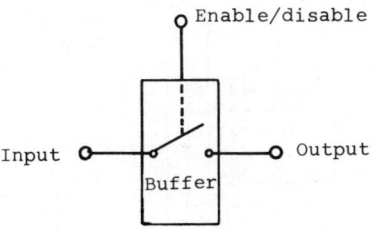

Fig. 12.3 Three-state buffer

Adders

These take two binary inputs and produce an output which is the sum of these in binary form. This is shown in fig. 12.4.

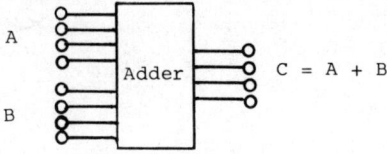

Fig. 12.4 Binary adder

Code converters

These devices are used to convert from one code to another. They can be used to convert binary-coded decimal (BCD) numbers to decimal for display on a seven-segment LED display unit, shown in fig. 12.5.

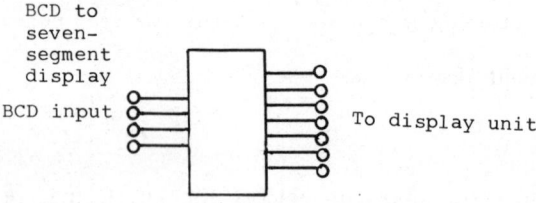

Fig. 12.5 BCD-to-seven-segment display

Multiplexers

Where several parallel inputs are required to be transmitted over a single line, a device called a multiplexer is needed to select one input at a time to transmit its data to the output. This process is known as multiplexing, and the reverse process is de-multiplexing. This is shown in fig. 12.6.

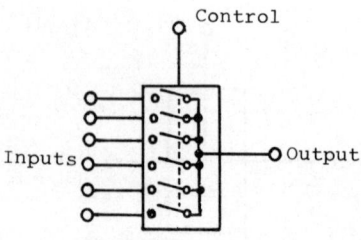

Fig. 12.6 Multiplexing

Memory blocks

Digital systems often require the storage of large amounts of data, and this can be by magnetic core, magnetic tape, or magnetic disc. The first two are gradually being replaced by semiconductor memories using flip-flops or charge-storage devices. Disc storage in the form of the small flexible or 'floppy' disc is however a very popular means of storing large amounts of data.

Semiconductor memories fall into two distinct types: registers, which are small fast-access short-term memories usually made up from J–K flip-flops; and longer-term permanent-memory chips, e.g. RAM, ROM, EPROM, etc. (these may be 'permanent' only as long as power is applied). Types of memory are explained in chapter 16.

A RAM memory may be able to store only one bit (binary digit) in each of 1024 (1K) locations or may store eight parallel bits (a byte) in 1K locations. The actual number of bits able to be stored is continually being increased – the 'one megabit' memory is being talked of. These devices (RAM's and ROM's) need extensive 'address' circuitry, i.e. circuits to select particular locations in the memory.

The two basic types of memory are illustrated in fig. 12.7.

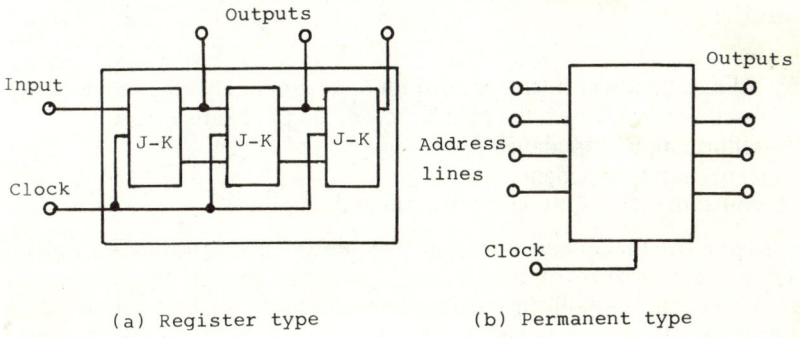

(a) Register type   (b) Permanent type

Fig. 12.7 Basic memories

13 Linear integrated circuits

In contrast to digital circuits, which are either on or off, linear or analogue integrated circuits can have an output of any value (within device limits) proportional to the input, and the output changes in a smooth manner as the input is changed.

These devices are normally used to increase – or amplify – current, voltage, or power.

General-purpose linear IC's

The circuits which perform the basic amplifier functions are known as operational amplifiers, or just op-amps, and are high-gain direct-coupled wideband amplifiers. Op-amps are assumed to have the following 'ideal' properties:

a) infinite gain;
b) infinite bandwidth (the bandwidth of an amplifier is the range of frequencies over which the gain is considered to be satisfactory);
c) infinite input impedance;
d) zero output impedance;
e) constant phase shift between input and output.

In practice the op-amp does not have these ideal characteristics, but the actual performance of the device is modified by the use of external components, normally using negative feedback, so that this is not a problem. The actual functions carried out may be straight amplification, comparison, voltage regulation, etc.

The op-amp is a complex network of transistors, diodes, resistors, and capacitors produced on a single chip by using integrated-circuit technology. These circuits could in practice be built using discrete components, but, unless such applications as power output and fast switching are required, the op-amp is cheaper, smaller, more reliable, etc.

Modern linear IC's have extremely complex internal circuitry. The internal circuitry for a relatively old and 'simple' IC – the 741 op-amp – is shown in fig. 13.1.

Access to the internal circuitry is not possible – the device comes in standard packages. These are shown in fig. 11.1, the most common being the eight-pin DIL (dual in-line) package. The op-amp symbol is shown in fig. 13.2. Note that it is the convention to omit power-supply pins on the symbol, since it is taken for granted that these must be available.

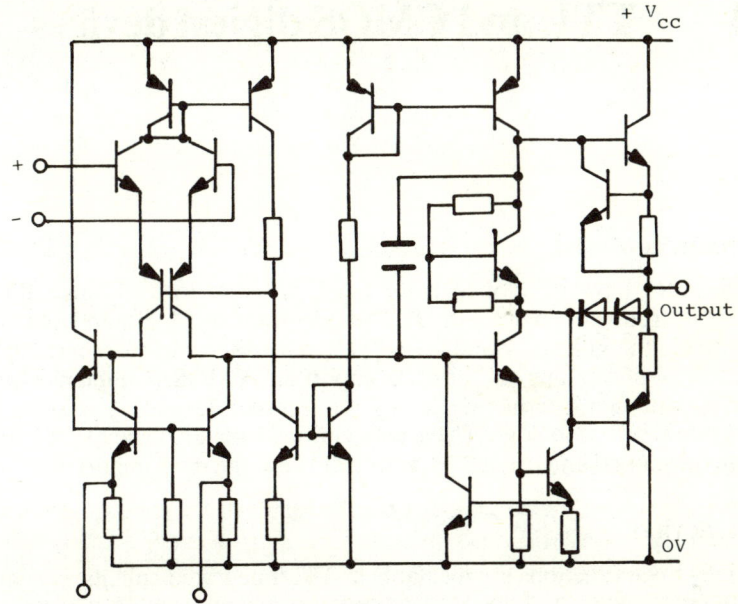

Fig. 13.1 Basic op-amp circuit

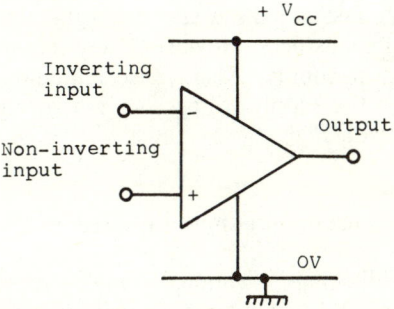

Fig. 13.2 Op-amp symbol

Interface circuits

An increasingly important aspect of modern electronics is the connection between the 'outside world' and the digital systems, referred to as 'interfacing'. This can also include the joining of incompatible digital devices.

Interfacing requires the using of specialist IC's such as analogue-to-digital (A-to-D) convertors, since most of the detector devices (transducers) and external signals are analogue. Finally conversion back from digital to analogue (D-to-A) is usually necessary.

14   TTL and CMOS digital devices

Introduction

Logic gates have developed from simple discrete logic through RTL (resistor–transistor logic) and DTL (diode–transistor logic) to what are effectively the industry standards: TTL (transistor–transistor logic) and, where low-power applications are important, CMOS (complementary metal–oxide semiconductor) logic. While other technologies such as integrated injection logic (I^2L) and emitter-coupled logic (ECL) are being developed and used, TTL and CMOS are the most important.

TTL devices

Various logic functions are available as TTL integrated circuits, ranging from simple gates and inverters to counters and shift registers (refer to any standard data book for the complete range). The device numbers and pin out connections of a range of popular devices are included at the end of the chapter.

TTL devices offer fast switching speeds, high-current output capability, and electrical robustness. However, they require a fairly closely controlled +5V power supply. Also, noise immunity – the ability to tolerate variations in the supply voltage – is not very good, and power consumption is relatively high.

The TTL NAND gate

The TTL NAND gate uses multi-emitter transistors. A simplified circuit is shown in fig. 14.1.

TTL has the following advantages when compared to other commercially available NAND devices:

a) Less area per gate (permits more devices per chip).
b) Faster operating speed.
c) As the output transistors, T_3 and T_4 in fig. 14.1, are in what is known as a 'totem-pole' configuration, with one transistor above the other, the device can provide or 'source' an output current or accept or 'sink' a current. This gives low-impedance drive in high and low output conditions, giving high speed and the ability to drive capacitive loads.

Referring to fig. 14.1, if either or both inputs A and B are connected to ground, i.e. logic 0, transistor T_1 will switch on, this will switch T_2, which in turn switches T_3 on and T_4 off; thus the output is connected through T_3 to V_{cc} or logic 1.

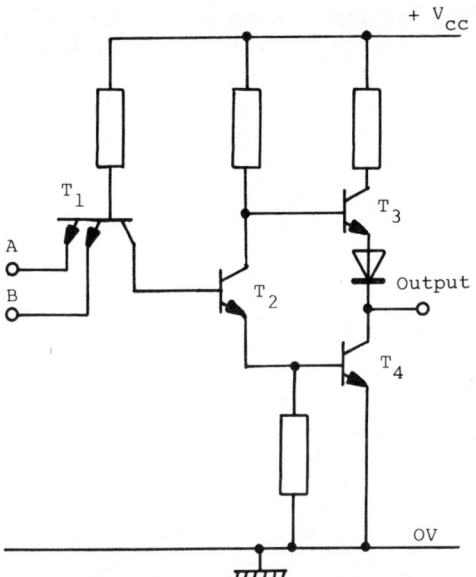

Fig. 14.1 The TTL gate

If *both* inputs are connected to logic 1, i.e. $+V_{cc}$ (5V), T_1 is held off, switching T_2 on, which switches T_3 off and T_4 on; thus the output is connected to ground (logic 0) via T_4. Thus the device carries out the NAND function.

An important fact can be observed from the above: to obtain a 0 output, all the inputs must be connected to logic 1 to switch transistor T_1 off; however, if the inputs are simply disconnected or left 'floating', T_1 will act as if it has a logic 1 on it. In practice, however, this is not a recommended technique – unused inputs should be connected through a 'pull-up' resistor to the supply voltage or be connected to another input.

Where it is necessary to connect the outputs of two TTL gates together, referred to as 'wire-OR' or 'wire-AND', the basic gate of fig. 14.1 cannot be used, since the 'totem-pole' output configuration ensures that if one input to one gate is high while one input to the other gate is low it is possible for the output transistors to draw excessive current and consequently be damaged. To overcome this, 'open-collector' gates with the upper transistor, T_3 in fig. 14.1, omitted are used.

Ranges of TTL
Various ranges of TTL are available. These include standard, low-power, high-speed, Schottky, and low-power Schottky devices. (A Schottky

diode consists of a metal-to-semiconductor (p-type or n-type) junction. However the Schottky diode has no minority carriers, which means that a Schottky diode can be turned off faster than a conventional diode. Also, the forward voltage drop across the Schottky diode when conducting is about half that for a silicon diode. A Schottky diode is connected between the base and collector of the input transistor to prevent it becoming saturated, thus speeding up operation.) The choice of a particular range depends on a number of factors, since in practice the choice is a compromise between speed and power, as can be seen by referring to fig. 14.2.

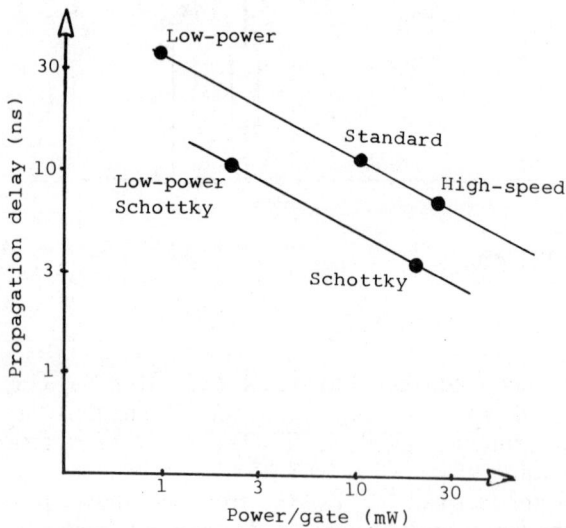

Fig. 14.2 TTL power/speed characteristics

Device types (recognition symbols)

TTL is available in a variety of packages and temperature ranges. A device code is used to indicate the basic function of the device and can be used to give other details.

The basic range can be prefixed by either 54 or 74.

Series 54: $-55°C$ to $+125°C$; military ranges; supply voltage $5\,V \pm \frac{1}{2}\,V$
Series 74: $0°C$ to $70°C$; industrial ranges; supply voltage $5\,V \pm \frac{1}{4}\,V$

Some typical codes are explained below.

a) 7400 N

        └──→ Package style.
               N = plastics dual in-line (DIL) package
               F = ceramic DIL package
       └──→ Device number, quadruple (or just 'quad') two-input NAND gate; i.e. there are 4 two-input gates per package. (Similarly, 'hex' means 6 gates per package.)

b) 74 H00 Again a quad two-input NAND gate, the H in this case indicating a high-speed device.

c) 74 S00 The S in the case indicates a Schottky device.

d) 74 LS00 The LS indicates a low-power Schottky device.

Maximum ratings

Supply voltage $V_{cc} = 7.0\,V$

Input voltage $V_{in} = 5.5\,V$

There may be electrical breakdown if V_{cc} exceeds 7 V. Logic operation is *not* guaranteed at all voltages below 7 V.

Data sheets

From the practical point of view, the most useful data is the pin out connections. For the basic TTL series these are included at the end of the chapter. Further details on the exact function of any device must be obtained from a complete data book.

Fan-out

This is the number of standard loads, each usually another input of the same family, which the device can supply or drive. 74-series devices usually have a fan-out of 10.

CMOS devices

CMOS integrated circuits are based on complementary pairs of field-effect transistors (p-type and n-type MOSFET's).

CMOS logic circuits have the advantages of low power consumption, wide supply-voltage range, and high noise immunity; but they suffer from such disadvantages as low output-current drive capability and susceptibility to damage by static charges during handling. Propagation delays (the time for a signal to pass through a gate) are also longer than for TTL.

Basic principles of CMOS devices

A simple CMOS inverter is shown in fig. 14.3. It consists of a p-channel MOSFET and an n-channel MOSFET. When the input voltage is high, the n-channel FET is turned on and the p-channel FET is turned off, so the output is low. When the input is low, the p-channel FET is turned on and the n-channel is turned off, so the output is high. Since the devices are FET's they do not exhibit saturation voltages or base–emitter voltage

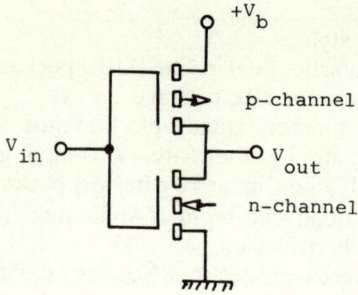

Fig. 14.3 Basic CMOS inverter gate

drops, so the output high and low states with no load are equal to $-V_b$ and zero respectively. This is shown by the transfer characteristic of the CMOS inverter in fig. 14.4.

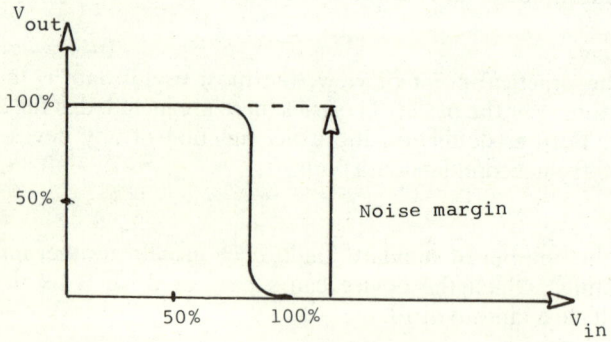

Fig. 14.4 Transfer characteristic of the CMOS inverter

From the diagram, we can see that the output will remain in the high or low states for quite large variations of the input high and low states. This means that CMOS devices have quite a large noise margin. It is important to note that open-circuit CMOS inputs do not float at logic 1, and should never be left floating.

Since one or other of the FET's in the complementary pair is always cut-off, the d.c. current consumption of a CMOS gate with no load is extremely small, so the power consumption is very low. Also, CMOS will operate over a wide range of supply voltages, since the lowest usable supply voltage is not limited by diode and base–emitter voltage drops, as with TTL, and increasing the supply voltage does not cause a large increase in current consumption. CMOS devices will operate quite well over a supply-voltage range of 3 V to 15 V.

Since FET's in the on state have a drain–source resistance of several kilohms, the output current that a CMOS device can source or sink is fairly limited. Manufacturers usually quote a current for which the output high voltage has fallen to a certain value or the output low voltage has risen to certain value; e.g. with a 10V supply a CMOS gate might be able to source 1mA before the high output voltage would drop below 9.5V and might be able to sink 2mA before the low output voltage would rise above 0.5V.

When interfacing CMOS with other circuits, the low current-handling capability can be a problem, and some sort of buffer device may be required. CMOS devices can however drive other CMOS devices adequately, since the input of a CMOS device is simply connected to the gates of two MOSFET's whose input resistance is very high, thus the input current is very low ($\simeq 10 \times 10^{-12}$A). Because of this, the fan-out is limited only by a.c. considerations and is normally limited by switching-speed requirements and for practical purposes is 50.

Handling CMOS devices

Since the outputs of a CMOS device are connected to the gates of MOSFET's and are insulated from the rest of the device only by an extremely thin layer of silicon diode, it requires a potential of only 50V to break down this oxide layer and thus destroy the device. Since static charges of up to 50kV can build up on objects insulated from ground (e.g. the human body walking across carpet can easily generate several kilovolts), internal protection is provided by the manufacturers; but this is not 100%, so certain handling precautions are necessary. These are listed below:

a) CMOS devices should always be supplied with their pins embedded in foil or conductive-plastics foam (*not* ordinary foam or expanded polystyrene) or an aluminium carrier. They should not be removed from the packaging until they are to be used. Care should be taken not to touch the pins after the device has been removed from the packaging and before it is inserted in the printed-circuit board or holder.
b) CMOS devices should always be the last components to be inserted into a circuit, and they should never be inserted or removed with power on.
c) If the devices are to be soldered directly into a circuit, then a low-leakage earthed soldering iron should be used. However, integrated-circuit sockets are to be preferred in general use.

Further information about the function of each type of CMOS device can be obtained from manufacturers' data sheets in a similar fashion to that for TTL devices.

The pin out connections of CMOS devices are also available on data sheets similar to the TTL data sheets on pages 89 to 94. Full details can be obtained from manufacturers' specifications.

Logic levels and noise immunity

The nominal voltage for logic 0 is usually 0 V, and logic 1 is $+V_b$ (5 V for TTL). In practice, this will not be the case.

In a practical logic family there will be small variations in the voltages representing the logic levels, and these must be taken into account. It is necessary to define how much the logic-0 voltage at an input is allowed to rise above the nominal level and how much the logic-1 voltage is allowed to fall below the nominal level without circuit malfunction. These levels are usually designated as $V_{il(max)}$ and $V_{ih(min)}$. It is also necessary to define the maximum permitted logic-0 output voltage and the minimum permitted logic-1 output voltage, which are designated $V_{ol(max)}$ and $V_{oh(min)}$ respectively. Clearly $V_{oh(min)}$ must be lower than $V_{il(max)}$. This is shown in fig. 14.5.

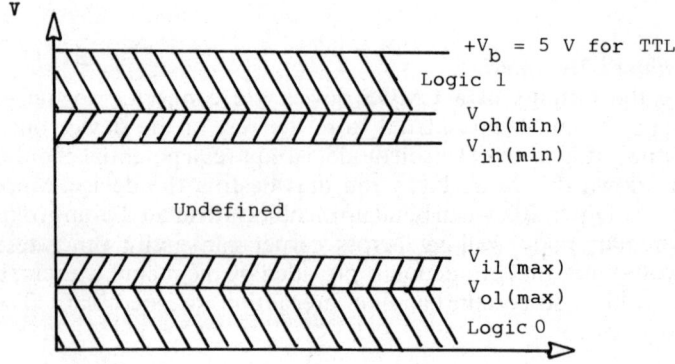

Fig. 14.5 Logic levels

Between these levels the output state of a logic gate is not defined.

Noise can appear on the inputs of logic circuits from a variety of sources. Current pulses from switching of other circuits can cause voltage transients on the power-supply lines, and transients on the mains-supply lines can appear on the power supply. Any noise pulse which takes the input of a logic circuit above $V_{il(max)}$ or below $V_{ih(min)}$ is likely to cause a malfunction.

Clearly the amplitude of noise pulse necessary to take the input into the undefined state is $V_{il(max)} - V_{ol(max)}$ for the logic-0 state and $V_{oh(min)} - V_{ih(min)}$ for the logic-1 state. These figures are known respectively as the low-level and high-level d.c. noise margins.

However, the d.c. noise margins do not tell the whole story, since they take no account of the duration of a noise pulse nor of the impedance of the logic circuits. If a noise pulse is long compared to the time a signal takes to pass through a gate (the propagation delay), then its amplitude will only need to be slightly above or below the d.c. noise margin to cause a malfunction. For shorter pulses a much larger amplitude is required to cause a circuit malfunction.

If the noise amplitude and pulse width are both taken into account then a figure known as the a.c. noise immunity is obtained.

TTL 74-series pin connections

Devices in the range of TTL (transistor–transistor logic) integrated circuits operating from a single supply of 5 V d.c. are capable of switching at frequencies generally in excess of 20 MHz.

Functions include gates, flip-flops, shift registers, decoders, counters, etc. All types are housed in standard DIL plastics packages and are suitable for operation over the temperature range 0°C to +70°C.

Logic thresholds for TTL are

'high' logic 1   2 to 5 volts
'low' logic 0   0.8 to 0 volts

Open-circuit inputs assume a 'high' logic-1 level.

The following abbreviations are used throughout the diagrams on pages 90 to 94:

A, B, C, D, and E	Data inputs binary weight (where applicable) $A = 1; B = 2; C = 4; D = 8; E = 16$
a, b, c, d, etc.	Segment outputs on seven-segment decoder driver
BCD	Binary-coded decimal
CP	Clock pulse
D, JK	Data inputs to flip-flops
GND	Ground 0 V terminal
LT	Lamp test
$Q, \bar{Q}$	Output and complement may have a letter indicating weighting (see above)
RBI	Ripple blanking input
RBO	Ripple blanking output
RC, C, R	Capacitor and resistor timing on monostables
V_{cc}	+ supply terminal
⫪	Schmitt device or function

Connections shown are top view.

A 'negation' circle at any output or input within the diagrams indicates that the terminal is active low or that at clocking inputs the device is negative-edge triggered.

7400 quad two-input NAND gate

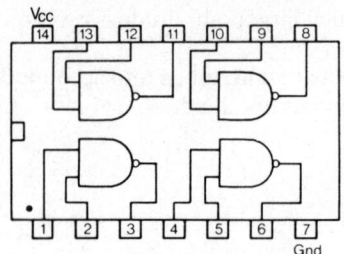

7401 quad two-input NAND gate with open-collector output

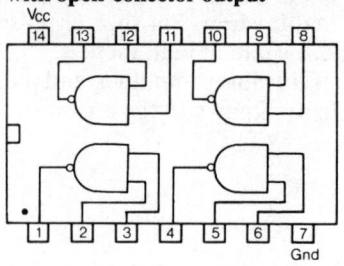

7402 quad two-input NOR gate

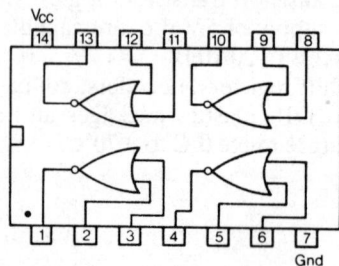

7403 quad two-input NAND gate with open-collector output

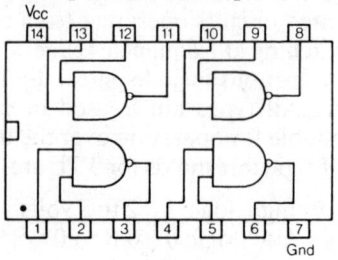

7404 hex inverter

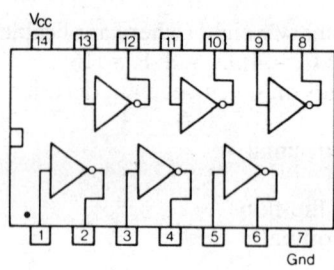

7406 hex inverter with open-collector output

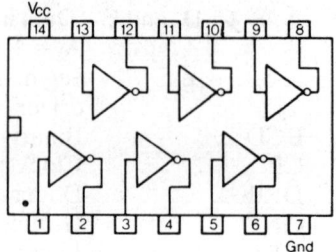

7407 hex driver with open-collector output

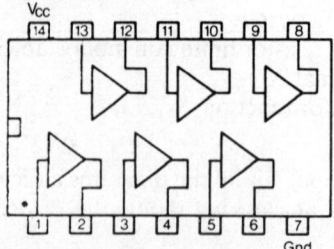

7408 quad two-input AND gate

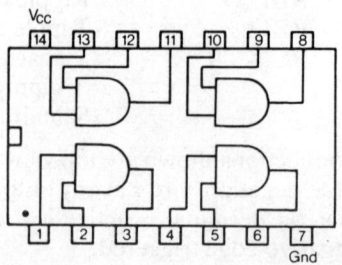

7410 triple three-input NAND gate

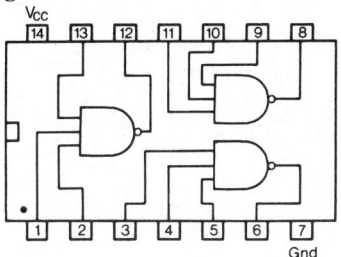

7413 dual four-input NAND-gate Schmitt trigger

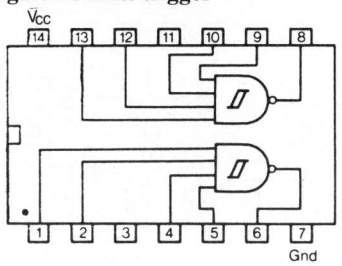

7414 hex Schmitt trigger

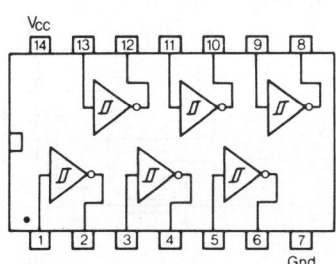

7416 hex inverter with open-collector output

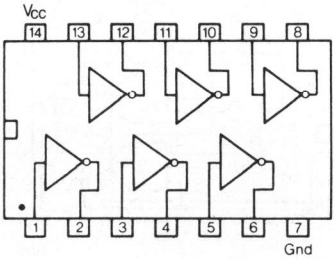

7420 dual four-input NAND gate

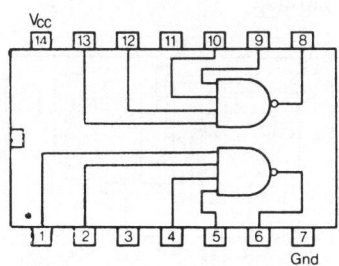

7430 eight-input NAND gate

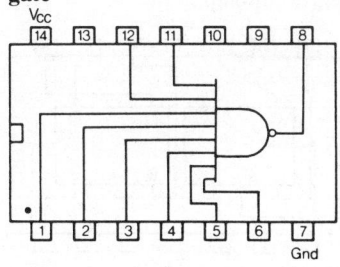

7440 dual four-input NAND buffer

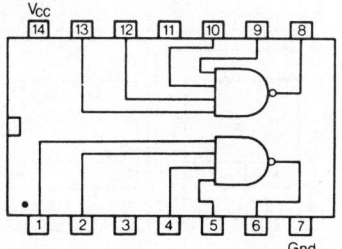

7442 BCD-to-decimal decoder

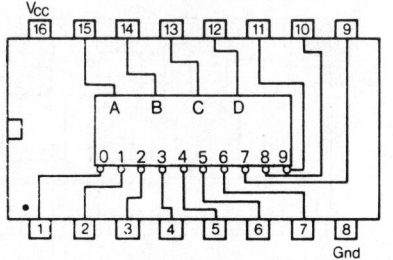

7447 BCD-to-seven-segment decoder–driver

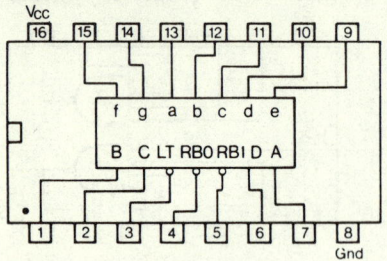

7451 dual two-wide two-input AND–OR–INVERT gates

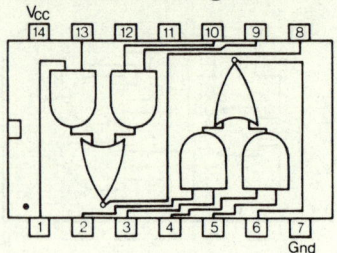

7470 J–K flip-flop

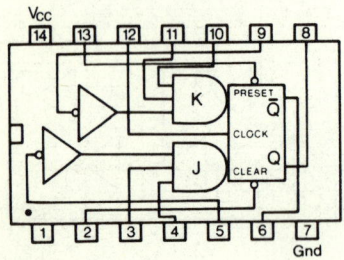

7472 J–K master–slave flip-flop

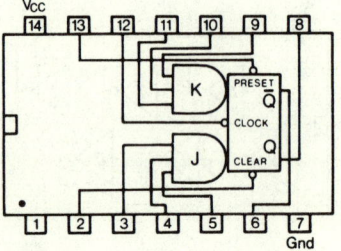

7473 dual J–K master–slave flip-flop

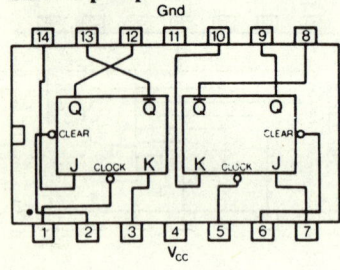

7474 dual D-type edge-triggered flip-flop

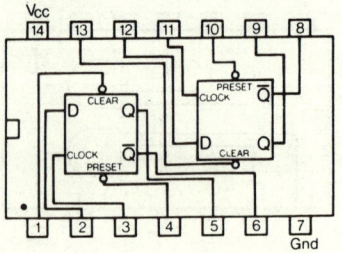

7475 quad bistable latch

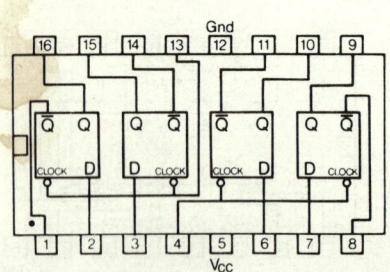

7476 dual J–K master–slave flip-flop with pre-set and clear

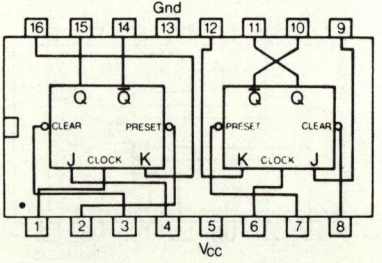

7486 quad two-input exclusive-OR gate

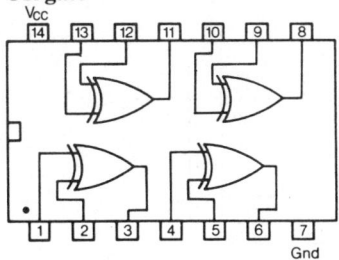

7490 decade counter

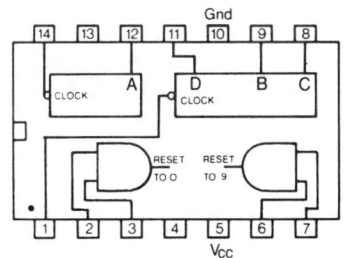

7493 four-bit binary counter

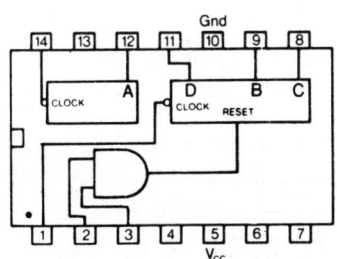

7496 five-bit shift register

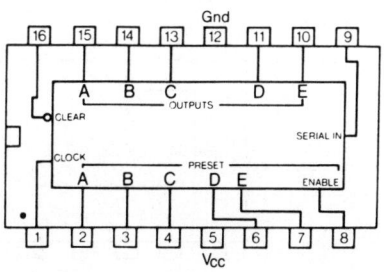

74107 dual J–K master–slave flip-flop with clear

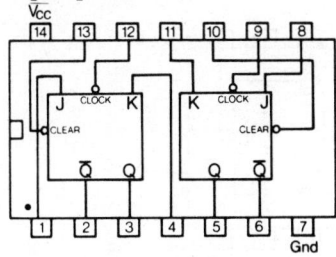

74121 monostable multivibrator

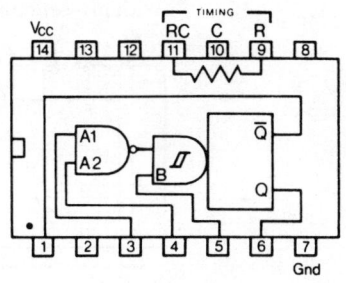

74123 monostable, retriggerable

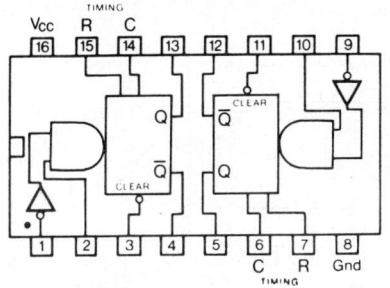

74132 quad Schmitt trigger

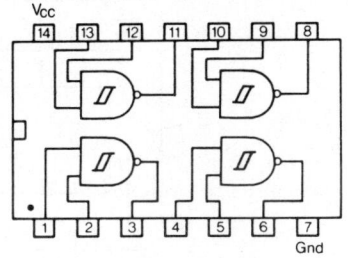

74141 BCD-to-decimal decoder–driver

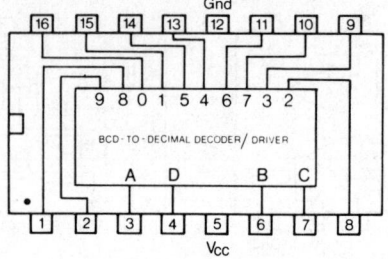

74154 four-to-sixteen-line decoder

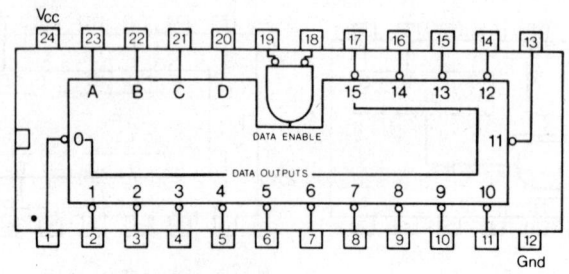

74192 up/down decade counter with pre-set inputs

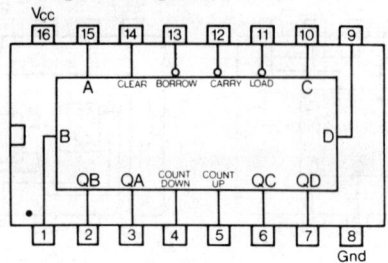

74193 up/down binary counter with pre-set inputs

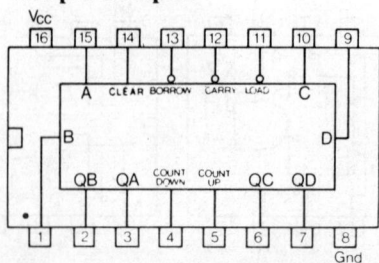

15   Practical logic systems

Introduction

The aim of this chapter is to introduce the use of the most common digital IC's in industrial equipment.

The most important examples at present are

a) transistor–transistor logic (TTL) – the 7400 series,
b) complementary metal–oxide semiconductor (CMOS) logic – the 4000 series.

There are variations on these examples, but the choice of configuration does not affect the general nature of the logic systems to be considered. For this reason the discussion will be confined to TTL, the differences between TTL and CMOS devices having been considered earlier.

Use of TTL devices

Type 7400 (quad two-input NAND gate)
This is a commonly used device since not only can it be widely used in its logic form, it can also be used as

a) a NOT element or inverter,
b) a buffer to external-signal input levels,
c) a set–reset latch operated by a 0 or 1,
d) a pulse generator.

These are the more obvious examples and are shown in fig. 15.1.

Type 7401 (quad two-input NAND with open collectors)
This can be used for applications similar to those for the 7400 IC, but it has the facility of being able to operate at a voltage level other than +5 V in order to be compatible with other systems – see fig. 15.2.

As the 74** type numbers ascend, so variations in the basic logic functions are provided. Some have a buffer mode, which implies a higher output-current capability. Some others have a Schmitt-trigger mode, which means that they will accept inputs whose levels depart from those specified for normal TTL. See fig. 15.3.

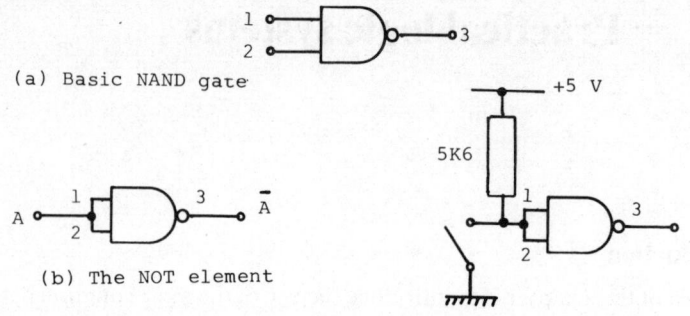

(a) Basic NAND gate

(b) The NOT element

(c) Input/logic interface

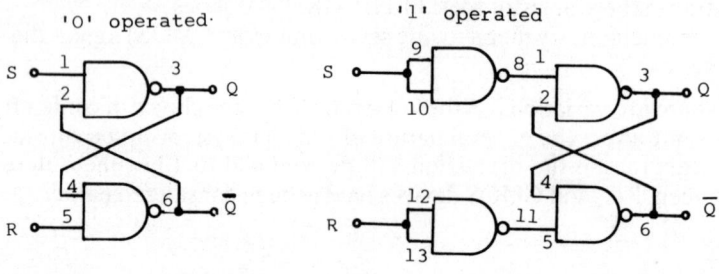

(d) Set-reset latch

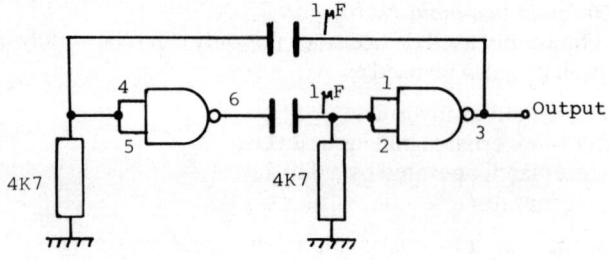

(e) Pulse generator

Fig. 15.1 Quad two-input applications

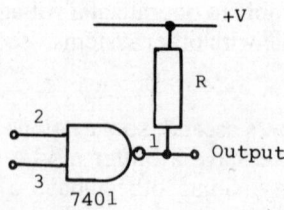

Fig. 15.2 Open-collector NAND gate

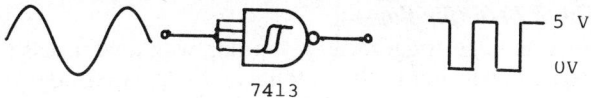

7413

Fig. 15.3 Schmitt-trigger NAND gate

Type 7442 (BCD-to-decimal decoder)

This will accept a four-digit binary code and, as the code progresses from 0000 to 1111, a 'low' will be available on each output in turn, designated 0 to 9. This IC may be used to select one from ten channels by means of a four-bit code. It may also be used to drive a cold-cathode numeric indicator tube – although other IC's are more suitable for this. See fig. 15.4.

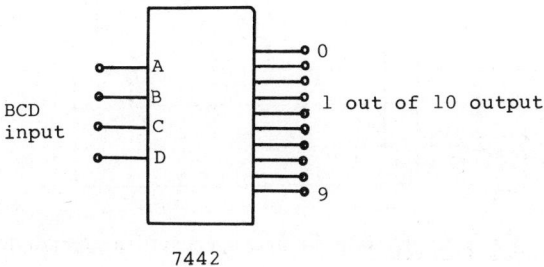

7442

Fig. 15.4 BCD-to-decimal decoder

Type 7447 (BCD-to-seven-segment decoder–driver)

This will accept a four-digit binary code and provide a combination of output signals from a total of seven which will energise the appropriate segments in a seven-segment LED display to provide a decimal representation of the applied BCD. See fig. 15.5.

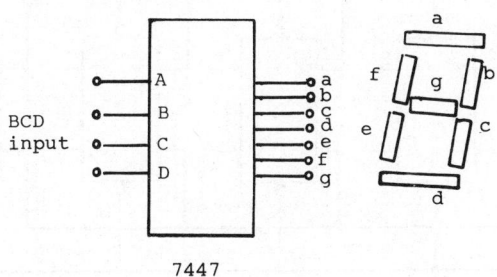

7447

Fig. 15.5 BCD-to-seven-segment decoder

Types 7470 to 7476 (flip-flops)

Type 7470 is an edge-triggered J–K flip-flop with dual inputs on J and K. Input data is transferred to the outputs on the positive edge of the clock pulse.

Types 7472, 7473, and 7476 are master–slave J–K flip-flops and are readily used in counters and shift-registers. 7472 and 7476 have a pre-set and clear facility which allows data to be pre-set in an arrangement of flip-flops or to be cleared from the arrangement. See fig. 15.6.

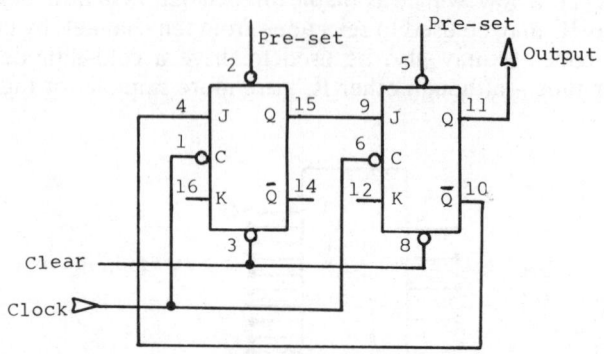

Fig. 15.6 Dual J–K flip-flop (e.g. 7476) connected for a count of three

Type 7474 (D-type flip-flop, edge-triggered)

The data input present on input D on the positive edge of the clock pulse is transferred to output Q. After the occurrence of this edge, any change on D is locked out. An application of a number of these elements would be to staticise digital data (i.e. to set the value – often the initial one – or status of a memory) before decoding and display. See fig. 15.7.

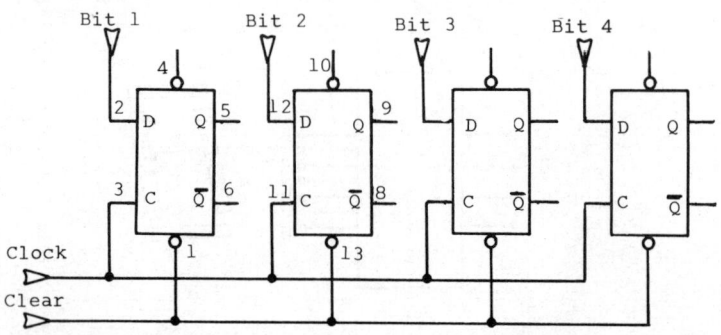

Fig. 15.7 Dual D-type flip-flop (e.g. 7474). Two IC's connected as a register

98

Type 7475 (quad bistable latch)
Data present on input D will be transferred to output Q when the clock pulse goes high, and Q will follow D as long as the clock pulse remains high. When the clock pulse goes low, the state of D in that instance is held on Q until the clock pulse goes high again.

Type 7486 (quad exclusive-OR gate)
The exclusive-OR gate will provide a 1 output only if inputs *A* and *B* are at different levels. The gate is useful in data comparison, arithmetic operations, code conversion, and error detection. See fig. 15.8.

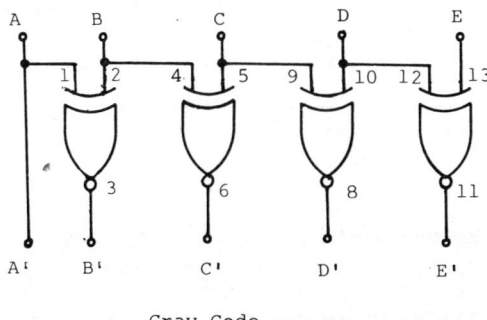

Gray Code

Fig. 15.8 Quad exclusive-OR (7486) connected as a binary/Gray converter

Type 7490 (decade counter)
This in fact consists of separate divide-by-two and divide-by-five counters which can be connected as a BCD counter giving an 8421-BCD output (see chapter 17) or as a divide-by-ten frequency divider giving a symmetrical output. See fig. 15.9.

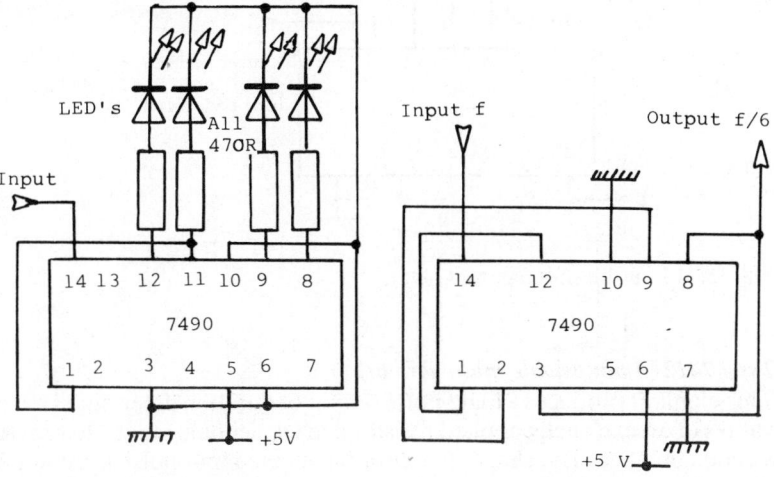

Fig. 15.9 Decade counter (7490)

Type 7493 (four-bit binary counter)

This in fact consists of separate divide-by-two and divide-by-eight counters which may be directly connected to give divide-by-sixteen (a pure binary count). See fig. 15.10.

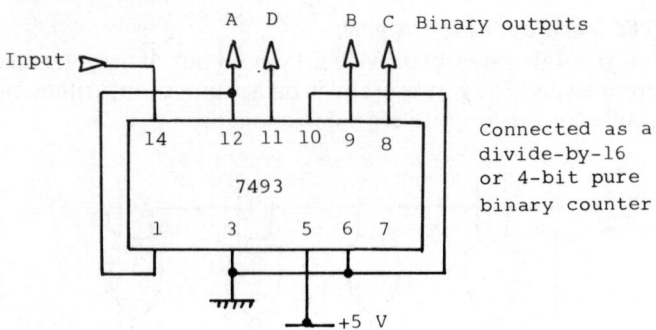

Fig. 15.10 Binary counter (7493)

Types 7494 to 7496 (shift registers)

Types 7494 and 7495 are five-bit registers, while 7496 is a six-bit register. In general, they may be used for serial-in (each digit must be presented one at a time in strict sequence), serial-out, and parallel/serial conversion of data etc. See fig. 15.11.

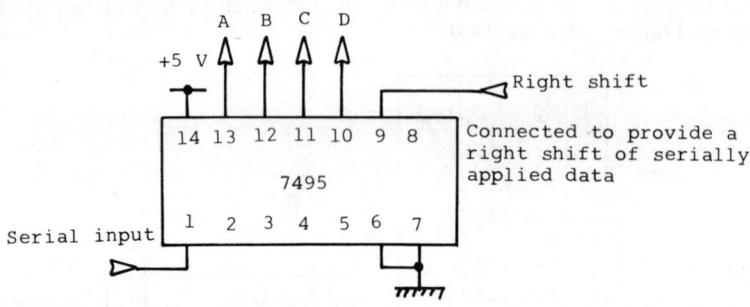

Fig. 15.11 Four-bit shift register (7495)

Type 74121 (monostable multivibrator)

This element produces a logic-pulse output of duration determined by the values of an externally applied resistor and capacitor arranged to provide a time constant. The shape and duration of the input pulse need not be defined. See fig. 15.12.

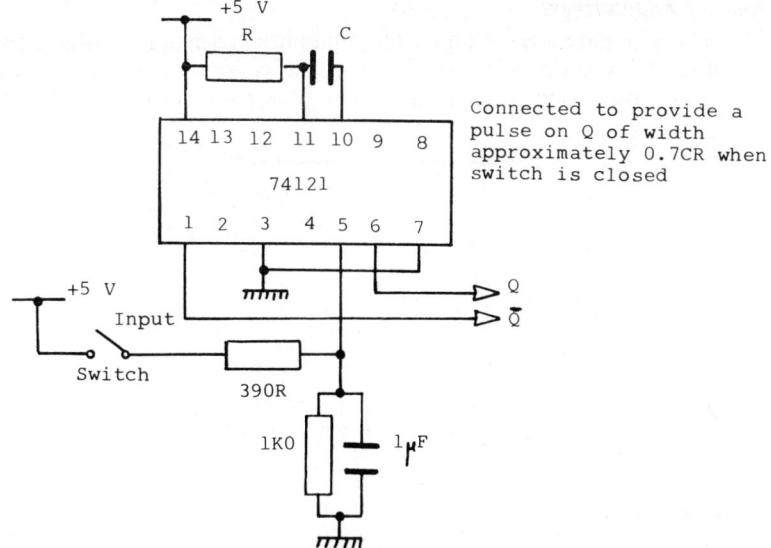

Connected to provide a
pulse on Q of width
approximately 0.7CR when
switch is closed

Fig. 15.12 Monostable multivibrator (74121)

Type 74154 (four-bit to sixteen-bit decoder)

This provides a separate single-bit output for each combination of a
four-bit binary code. For example we can select any one of sixteen
channels by means of a four-bit address.

More applications

Contact debounce

Where a switch or relay contact is required to provide an input level to a
logic system, any contact bounce may give rise to a succession of pulses
which may, in error, be treated as valid inputs and cause a malfunction.
Figure 15.13 shows how a set–reset element may be used to transmit the
initial contact only.

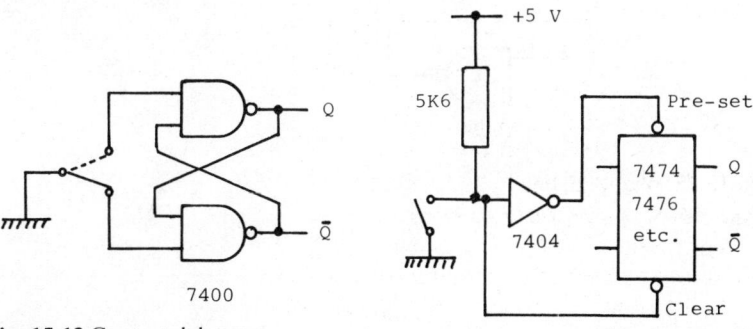

Fig. 15.13 Contact debounce

Lamp or relay drivers

The basic TTL gate does not have the capability to driver more than, say, a light-emitting diode (LED). We can, however, use open-collector devices in conjunction with a transistor amplifier as shown in fig. 15.14.

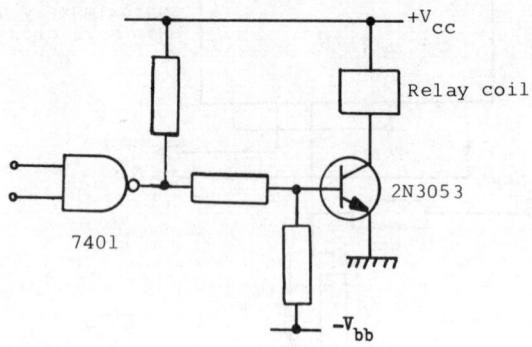

Fig. 15.14 Driver circuit

Wired-AND

This is the technique of directly connecting logic-element outputs together in order to realise a particular required logic function. This is not recommended for standard TTL outputs, but is one of the reasons for the open-collector output. Figure 15.15 shows how two gates may be 'wire-ANDed'.

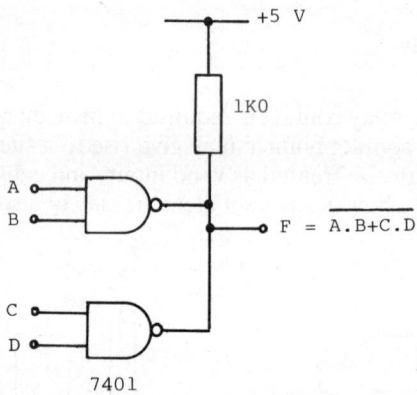

Fig. 15.15 'Wire-ANDing'

16   Programmable controllers

Introduction

Industrial-control systems have traditionally been based on hard-wired relay or equivalent solid-state equipment which by nature is bulky, inflexible, and subject to soaring costs.

The advent of digital techniques and cheap integrated circuits has encouraged the programmable approach, so that a given system is able to cope readily with variations without it being necessary to rewire and replace equipment.

The principle of operation of a programmable controller is not far removed from that of a general-purpose computer, the main difference being that the former does not need the processing capability of the computer, so its functions and software requirements are much simpler. For example there are no levels of priority in a programmable controller.

Description

The programmable logic controller (PLC) is required to perform a sequence of operations by first scanning inputs such as relay contacts, limit switches, push-buttons, etc. and then comparing the status of these devices with conditions specified in a program. Output devices such as contactors, stepper motors, solenoids, etc. are then energised or de-energised as required.

Figure 16.1 shows the basic block diagram of a typical PLC.

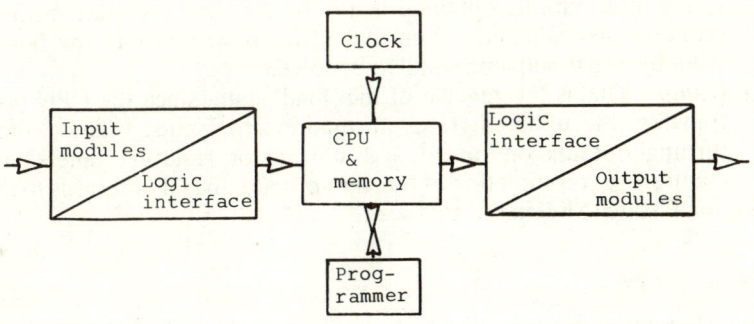

Fig. 16.1 Basic block diagram of a typical PLC

The central processor unit (CPU) contains a working memory (RAM, explained later in this chapter) and has access to a memory which contains programs relating to the sequences to be carried out. It performs the required logic functions and contains, or has control of, timers, counters, shift registers, and latches.

The clock and the timing circuitry synchronise all logic operations performed by the CPU.

The programmer is the means by which the programme is loaded into the memory of the PLC. It may take the form of a keyboard, a teletypewriter, a visual display unit, or cassette magnetic tape. It may also be used to monitor the status of the PLC for fault-finding etc. A programmer may not be a permanent feature of any particular PLC but may be plugged in as required.

Input and output modules are the means by which a wide range of electrical quantities may be safely accommodated and converted to appropriate logic levels by the logic interface equipment.

The central processor unit (discussed in greater detail in chapter 17)

The main components of the CPU may be identified as

a) the data buffer store (a temporary data store),
b) the logic unit.

Each scan of the CPU may be divided into three states, as follows:

i) *Load* The CPU sequentially addresses the input/output sections and transfers the status of each input, output, counter, timer, etc. into a corresponding address in the data buffer store. When this task is complete, the CPU moves on to the next task.

ii) *Solve* During this state, the CPU addresses the memory and executes the program instructions in sequence in accordance with the status information contained in the buffer store. As each control sequence is evaluated, its current status is written into the buffer store for use in sequences still to be solved.

iii) *Dump* This is the reverse of the 'load' state, since the CPU now transfers the updated status information to inputs, outputs, etc., turning outputs on or off and starting or resetting timers and counters. On completion of the 'dump' state, the CPU returns to the 'load' state to begin the next scan.

The memory

Nowadays memories are solid-state whereas, in the past, core stores and plated-wire stores have been used. Solid-state memories fall into two main categories:

i) random-access memory (RAM),
ii) read-only memory (ROM).

The random-access memory has a convenient read/write capability but is volatile – i.e. the contents are lost if there is an interruption in the supply. However, RAM may be provided with battery back-up. Apart from the data buffer store, the RAM is useful where frequent program changes are expected or as an aid to program development or debugging.

The read-only memory implies that the contents are pre-loaded and are not changed during the operation of the PLC. Types of ROM are as follows.

a) Mask programmable.
b) Fused-link programmable (PROM). These devices are supplied with all memory locations set to 0. Selected locations can be changed to 1 by a procedure which involves using special equipment to drive a high current through the IC to open up a fusible link (similar to blowing a fuse). This is non-reversible.
c) Erasable programmable (EPROM).
d) Electrically alterable ROM (EAROM).

The ROM generally contains the program, and the PLC manufacturer usually provides the means of entering new programs or erasing old ones.

Programming

Traditionally, sequence-control systems are hard-wired in electro-magnetic-device circuitry. The manner in which such a system is documented for the purposes of reference is by way of a *ladder diagram*, an example of which is shown in fig. 16.2.

The sequence is as follows. Operating push-button PB_1 energises relay A which is latched into operation by its contacts A_1. In the next step, contacts A_2 are closed in preparation for the operation of limit switch LS_1 which energises relay B through the normally closed contacts C_1.

Contacts B_1 close and solenoid 1 is operated. Contacts A_2 will also prepare limit switch LS_2 for energising relay C. When C is operated, contacts C_1 will open to ensure relay B is released, and contacts C_2 will close to energise solenoid 2. The STOP will release all relays and solenoids at any time.

In a programmable logic controller, relays A, B, and C would not exist. Instead, inputs PB_1, STOP, LS_1, and LS_2 would be scanned for their logic status, and the state of solenoids 1 and 2 would be updated in accordance with the required sequence specified by the stored program. In most equipments, the style of programming and the symbols used are related to the traditional relay ladder diagram, as shown in fig. 16.3.

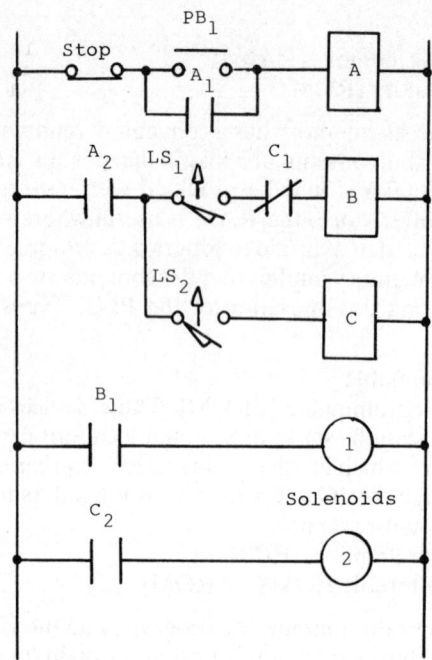

Fig. 16.2 A ladder diagram

Programming the given example

The ladder function symbols shown in fig. 16.3 are associated with many PLC's and may be present on a programmer keyboard. Alternatively, the logic functions STORE, AND, etc. may be typed in as required.

Considering our example, a PLC may require us to assign X_1, X_2, etc. to inputs and to assign Y_1, Y_2, etc. to outputs, so that we have

X_1 represents STOP
X_2 represents PB_1
X_3 represents LS_1
X_4 represents LS_2
Y_1 represents LATCH (usually an internal facility)
Y_2 represents solenoid 1
Y_3 represents solenoid 2

The sequence may start by examining the state of STOP, which should not be operated. Therefore its complement state – the opposite logic state – is required to start a logic line. We next need to check to see if PB_1 has been operated. If this is the case then a latch Y_1 is provided to hold the logic line in the TRUE condition until changed to FALSE when STOP is applied.

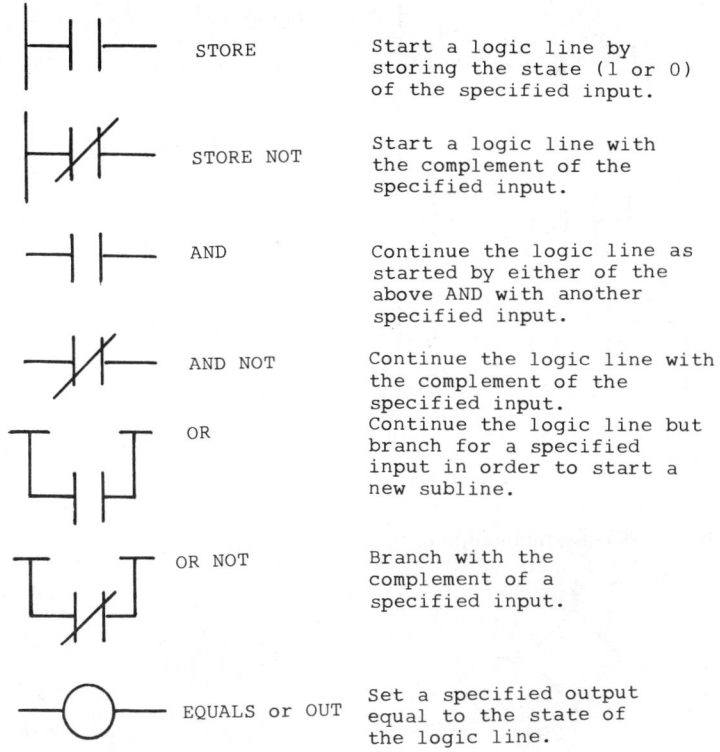

⊣⊢	STORE	Start a logic line by storing the state (1 or 0) of the specified input.
⊣/⊢	STORE NOT	Start a logic line with the complement of the specified input.
⊣⊢	AND	Continue the logic line as started by either of the above AND with another specified input.
⊣/⊢	AND NOT	Continue the logic line with the complement of the specified input.
	OR	Continue the logic line but branch for a specified input in order to start a new subline.
	OR NOT	Branch with the complement of a specified input.
─○─	EQUALS or OUT	Set a specified output equal to the state of the logic line.

Fig. 16.3 Ladder-diagram symbols and functions

When LS$_1$ is operated, solenoid 1 must be operated, provided that the latch Y$_1$ is in operation and LS$_2$ is not operated.

When LS$_2$ is operated, solenoid 2 must be operated and solenoid 1 must be released, provided that the circuit is still latched by Y$_1$.

When the STOP switch has been operated, the latch is released and all logic lines assume a FALSE state, causing the solenoids to be released.

The ladder solution is shown in fig. 16.4.

Boolean representation

The ladder-diagram representation might be very acceptable to those used to relay sequence diagrams but may be tedious to those who are prepared to accept the system sequence as an equation and enter it through a programmer as such.

The previous example may be expressed by Boolean equations as follows:

107

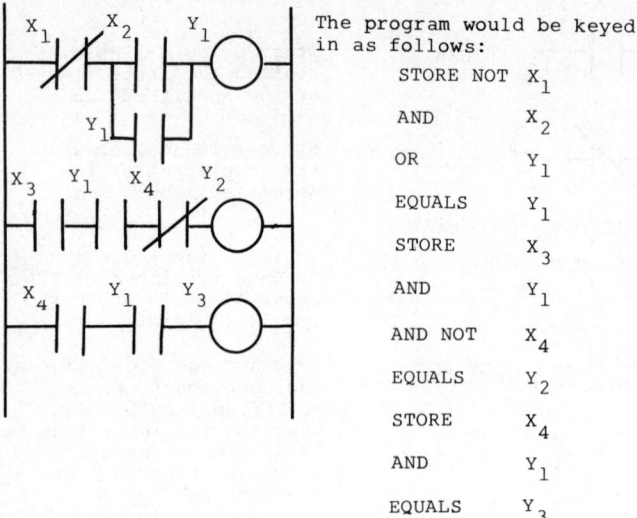

The program would be keyed in as follows:

STORE NOT	X_1
AND	X_2
OR	Y_1
EQUALS	Y_1
STORE	X_3
AND	Y_1
AND NOT	X_4
EQUALS	Y_2
STORE	X_4
AND	Y_1
EQUALS	Y_3

Fig. 16.4 Ladder-diagram solution

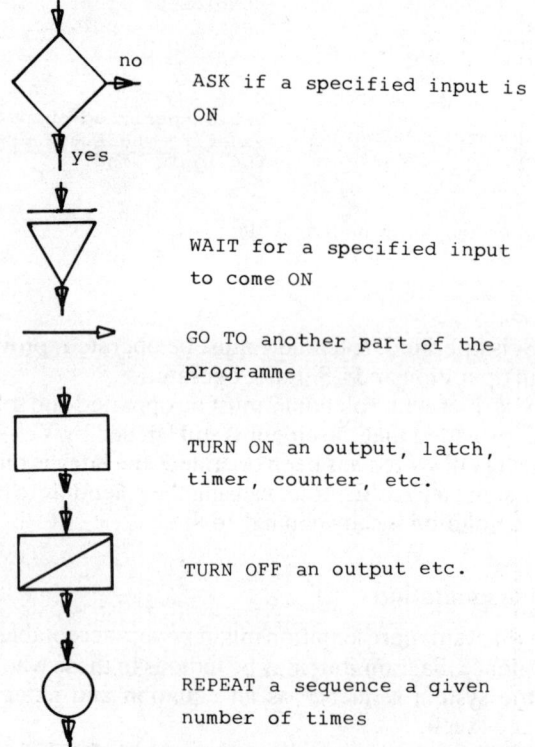

ASK if a specified input is ON

WAIT for a specified input to come ON

GO TO another part of the programme

TURN ON an output, latch, timer, counter, etc.

TURN OFF an output etc.

REPEAT a sequence a given number of times

Fig. 16.5 Typical flow-chart symbols and functions

$$\overline{X}_1 . (X_2 + Y_1) = Y_1$$
$$X_3 + Y_1 + X_4 = Y_2$$
$$X_4 + Y_1 = Y_3$$

where the logic symbols have the usual connotations.

Flow-chart representation

In some PLC's, flow-chart programming is provided, since compared with ladder or Boolean programming it economises on the steps needed in some sequences.

Typical flow-chart functions
The flow-chart for the previous example is shown in fig. 16.6.

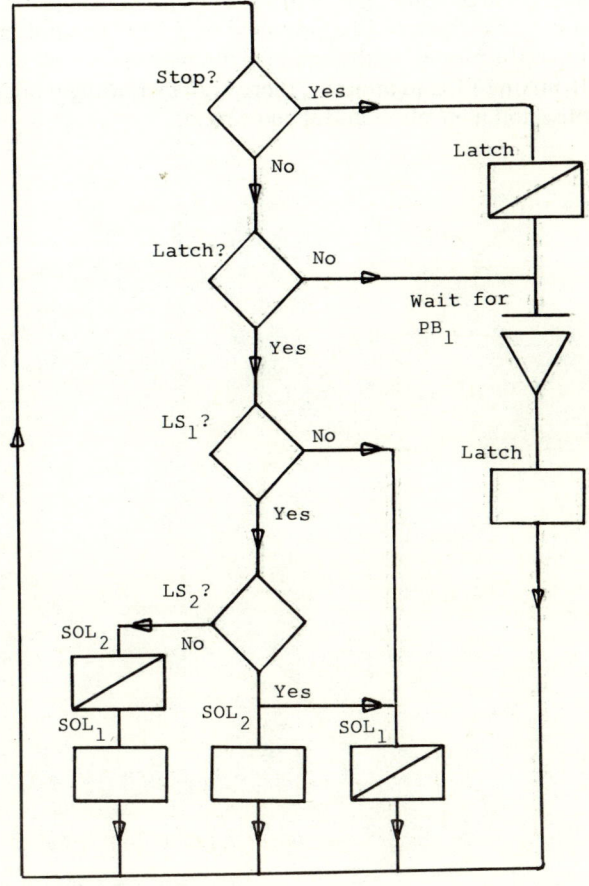

Fig. 16.6 Flow-chart solution to example

Flow-chart symbols are available on a keyboard so that the program may be entered directly with reference to an appropriate flow-chart.

Summary

The given example should illustrate that there need not be a strict relationship between timing in the system, as determined by push-switches and limit switches, and the scan time of the PLC – provided that the scan time is fast enough to produce the necessary system response. However, the PLC can be made to wait for a given input before it is allowed to proceed.

Suppose that in the example we had required a delay of say 10 seconds between the operation of LS_2 and solenoid 2. The PLC would have provided this if we had included TMR10 as a block in the logic line. Similarly we could have counted a number of operations by means of CTR and provided some action after a given count.

The functions which can be provided by a PLC depend on the manufacturer and the model. Although such functions may vary in number and nature from one PLC to another, there is an increasing tendency towards standardisation in symbols and specification.

17 Introduction to microprocessors

The microprocessor or MPU (microprocessor unit) – a single IC or a set of a few IC's which can be programmed with stored instructions to enable a wide variety of functions to be carried out – is a natural development of solid-state electronics, and is probably the most important. Using micro-miniaturisation techniques, manufacturers are able to incorporate many thousands of components on to a single silicon chip.

The microprocessor is designed to recognise and obey a set of instructions known as the 'instruction set'. By supplying the device with the instructions in a suitable order, the device can be made to carry out arithmetic and logic operations, data manipulation, and control functions.

A complete explanation of the internal operation of a microprocessor is beyond the scope of this book, and is in any case not directly relevant to the service technician, since he is presented with a working system and only the external signals of the device can be checked. In practice, due to the sophisticated nature and cost of the necessary test equipment, e.g. logic analysers, the servicing of complex digital systems will probably be carried out by specialist agencies. However, a brief look at the basic principles may be of interest.

The microprocessor device

The microprocessor – the 'computer on a chip' – is made up of three basic blocks of varying complexity and size. The heart of the processor is the central processor unit (CPU). This performs the actual operations and is controlled by a set of instructions supplied to it in sequence. An *instruction* is a control word containing enough information to enable the CPU to carry out one of its predetermined functions.

To enable the CPU to function, there must be some data-storage facility. At present this is separate from the microprocessor device itself, but it will probably be incorporated on the device chip itself as the manufacturing technology is improved. The data storage is required for two reasons:

a) to store data that is being worked on at any time and to store results of operations. These are normally stored in a random-access memory (RAM), which is a device or system which may have data written in or read out at high speed independent of its position in the store.

111

b) to store the instructions which control the operation of the CPU. These instructions are stored in read-only memory (ROM) or erasable programmable read-only memories (EPROM's), which are permanent or semi-permanent memories, i.e. memory devices or systems which have their data content written in and permanently held (or held until some specific erase operation is carried out).

The third block in the computer is the input/output (I/O) function. To carry out any useful function, the computer must be able to communicate with the 'outside world', i.e. data must be put into and taken from the memory.

The arrangement of these three main functions is shown in fig. 17.1.

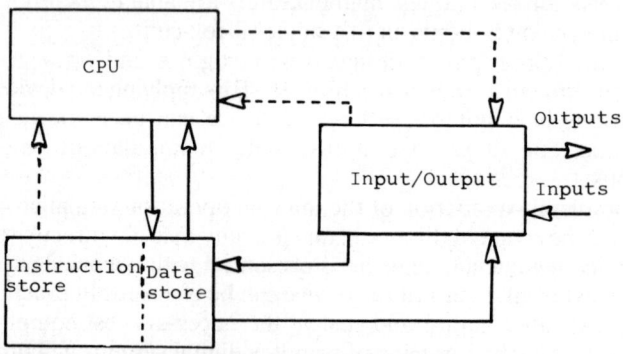

Fig. 17.1 Microprocessor block diagram

The instruction store, although part of the main memory, is often thought of as a separate memory.

The sequence of control instructions for a given task is called a *program*, and the writing of a program is called programming. A general term for programs and related information is *software*, as opposed to the term covering the actual devices in the system, which is *hardware*.

Binary number systems

The majority of microprocessing systems available are based on a binary word that is eight bits (binary digits) long, known as a *byte*.

To the microprocessor, a byte is merely a set of eight electrical signals, or logic levels. Each signal can have one of two possible values, or states.

The machine itself does not know whether these electrical states represent a binary or any other kind of number. The designer of the internal logic of the MPU gives particular bit patterns significance by the way he organises the logic.

While the MPU actually operates on binary signals, it is possible that other binary-based numerical scales may be met, though these are more

Table 17.1 Number systems

Decimal	Binary	Hexadecimal	Octal	BCD
0	0000	0	0	0000
1	0001	1	1	0001
2	0010	2	2	0010
3	0011	3	3	0011
4	0100	4	4	0100
5	0101	5	5	0101
6	0110	6	6	0110
7	0111	7	7	0111
8	1000	8	10	1000
9	1001	9	11	1001
10	1010	A	12	0001 0000
11	1011	B	13	0001 0001
12	1100	C	14	0001 0010
13	1101	D	15	0001 0011
14	1110	E	16	0001 0100
15	1111	F	17	0001 0101
16	10000	10	20	0001 0110

likely to be met by the engineer who programs the MPU. These systems include 'octal' and 'hexadecimal', with 'binary-coded decimal' (BCD) in many applications as a compromise.

Table 17.1 shows the four systems and equivalents.

The MPU operates most efficiently with binary number systems. However, there are many occasions when it is required to operate with decimal numbers, using BCD. BCD is similar to hexadecimal, except that the characters A to F are not used. Thus we can see that with a single byte it is possible to express a two-digit decimal number within the limits

0000 0000 (BCD) $\equiv$ 00 (decimal)

1001 1001 (BCD) $\equiv$ 99 (decimal)

The most common form of BCD is known as 8421-BCD. The name is derived from the fact that the decimal equivalent of a four-bit binary number is found by multiplying each binary digit by 1, 2, 4, or 8, depending on its position. For example,

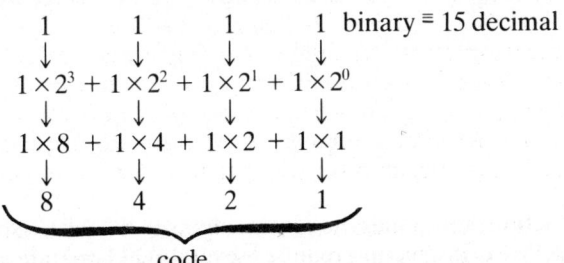

$$
\begin{array}{cccc}
1 & 1 & 1 & 1 \quad \text{binary} \equiv 15 \text{ decimal} \\
\downarrow & \downarrow & \downarrow & \downarrow \\
1 \times 2^3 + & 1 \times 2^2 + & 1 \times 2^1 + & 1 \times 2^0 \\
\downarrow & \downarrow & \downarrow & \downarrow \\
1 \times 8 + & 1 \times 4 + & 1 \times 2 + & 1 \times 1 \\
\downarrow & \downarrow & \downarrow & \downarrow \\
8 & 4 & 2 & 1
\end{array}
$$

code

Microprocessor hardware

MPU's produced by various manufacturers differ slightly in their internal layouts and instruction sets – these are the total lists of instructions which can be executed by given MPU's. However, the following simple explanation should explain some of the basics.

The section of the microcomputer that actually performs all the data manipulation is the central processor unit (CPU). It comprises a number of *registers* – small-scale memories storing words which may involve arithmetical, logical, or control instructions. This includes a good deal of control logic, and since this control logic is not directly accessible to the service engineer it is of no direct interest.

The MPU uses a 'program counter' – a register in the CPU – to hold the address of the next instruction. This address is a unique memory location for a statement which specifies the values or locations of the quantity or logic function (called the *operand*) which is to be carried out next, so that the machine can be 'stepped' through its program. The 'stack pointer' is a register used to co-ordinate the storing and retrieval of data in the stack – which is simply a block of successive memory locations. The 'index register' contains address data which can be altered by the control unit without affecting the data in the memory.

The program counter holds the address of the next instruction to be carried out. Once the instructions are stored in the required sequence in the memory, the processor can read these locations each time it requires a new instruction. They can be reached by presenting the program counter with the first memory address and automatically incrementing it by 1 after the execution of the instruction. It is possible to alter the program-counter contents to alter the program sequence to permit 'jumps' or repeats.

Once an instruction has been obtained from the memory, the MPU performs the operation defined by that instruction. One of the major blocks within the MPU is the *arithmetic logic unit* (ALU) which performs the arithmetic and logic functions. This circuit has two sets of data input and one of data output and performs arithmetic or logic operations on the two input words and gives the results at the output.

The actual function carried out by the ALU is determined by a set of control inputs. Some typical control functions are listed below.

AND	add
OR	subtract
NAND	invert
NOR	shift left
exclusive OR	shift right

The MPU instruction determines the control inputs to the ALU. Since, in general, a single MPU instruction requires several ALU operations, a device called a 'sequence controller' is used to decide on the correct sequence of operations.

For fast operation the CPU requires some internal working storage area to be available for immediate data storage without the necessity for addressing a memory location via a processor or instruction.

It is usual to have a 'status register' to indicate which instruction was last carried out.

The basic system described briefly above is shown in fig. 17.2.

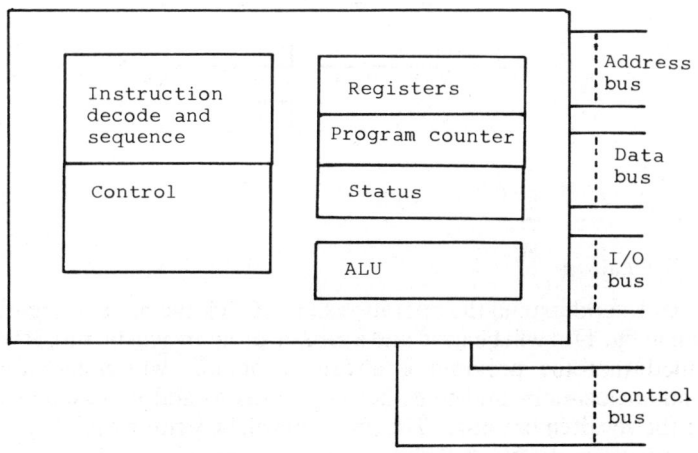

Fig. 17.2 CPU block diagram

The MPU must also communicate with the memory devices, input and output devices, etc., and it does this by means of 'busses', which are simply multi-wire 'highways'. These are referred to as the data bus, address bus, control bus, and input/output (I/O) bus. To reduce wiring complexity, some systems carry more than one of the above functions on a single bus, by using a suitable switching arrangement, e.g. 'tri-state buffers'.

The MPU is a device operating sequentially at a fixed rate. This rate is controlled by a generator of pulses known as the 'clock'. The clock generates several pulses spaced at different time intervals to ensure that the correct operation is performed at the right time, which will be determined by how long it takes to perform each instruction.

In the basic CPU shown in fig. 17.4, the clock produces seven pulses t_1 to t_7. These are time separated as shown, with the width of each clock pulse depending on the MPU. Whenever a clock pulse is applied to a gate, it 'enables' that gate to pass the data presented at its second input. The cycle of clock pulses shown in fig. 17.3 is repetitive.

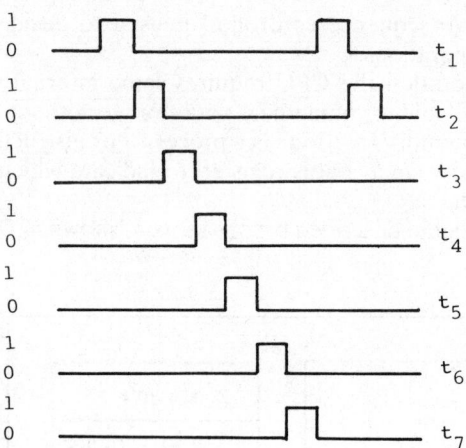

Fig. 17.3 Cycle of clock pulses

CPU operation

In order to understand the operation of the CPU, the basic configuration shown in fig. 17.4 will be used and a sample program will be run. It will be assumed that the program is already contained within the store in sequential locations and that the program is to add two numbers and place the result in the store. The program will be written as:

Store location	Instruction
010	Transfer the contents of location 060 into the adder.
011	Transfer the contents of location 137 into the adder.
012	Perform addition and place result in ACC.
013	Store the contents of ACC in location 047.

Assume location 060 contains 20_{10} and location 137 contains 16_{10}. (The subscript, or radix, 10 to the numbers 20 and 16 indicates that the numbers are decimal; e.g. 10_2 is to base 2 or a binary number.)

Step 1 – the fetch cycle

When this program is allowed to run, the program counter contains the instruction address 010. At time t_1 this is clocked into the address register, where it is then used by the decoder to open location 010. When the clock pulse t_2 occurs, it causes the contents of 010 (first instruction) to be fed into the data register.

Once assembled there, clock pulse t_3 causes the instruction to be fed into the instruction register.

116

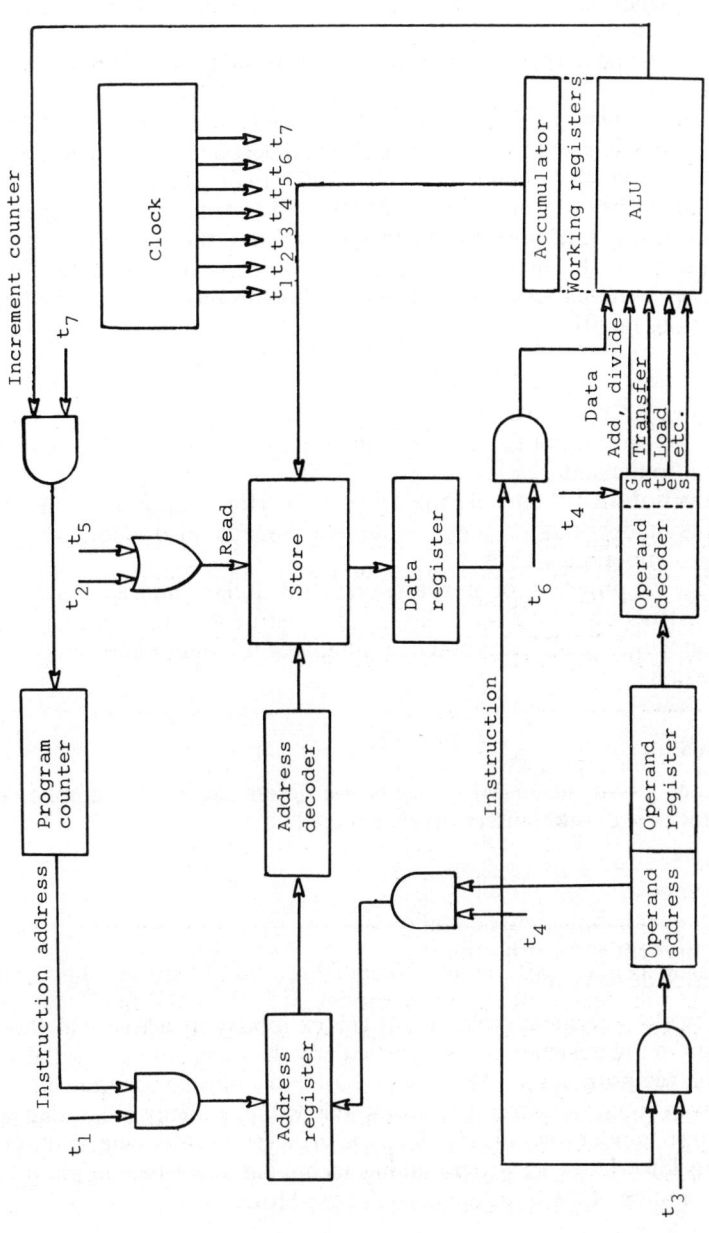

Fig. 17.4 Basic configuration of the CPU

117

Step 2 – the execute cycle

The operand-code portion of the instruction is fed into the operand decoder, where it is decoded and used to provide one input to a series of gates.

At time t_4 one of the gates is allowed to open and the instruction is set in the ALU.

Simultaneously the operand address (060) which is part of the instruction is sent to the address register and decoder, where it is used to open location 060 containing the first number to be added.

At time t_5 the contents of location 060 are then fed to the data register, and at t_6 this is transferred to the registers within the ALU. A signal is then applied on the increment lead to the increment counter, and at t_7 this counter is incremented by 1. Thus it now contains the address of the next instruction, i.e. 011.

The above fetch and execute cycles are performed and, according to the instruction stored in 011, the contents of location 137 (i.e. 16_{10}) will be placed into the adder.

Instruction 012 then tells the adder to perform addition using two numbers held within the adder. Note that the 'add' instruction does not contain an operand address.

Finally the program counter steps on to 013, and this address causes the store instruction to be read. Using this instruction, the result of the addition which is held within the accumulator is fed into location 047 of the address.

Conclusion

A system based on an MPU offers the design engineer a number of advantages over other alternatives, namely

a) smaller size,
b) lower cost,
c) greater reliability and flexibility,
d) component standardisation,
e) shorter design cycle.

Practically, this means that the MPU is extremely attractive to all those involved in the design process – particularly the user. Since, with the aid of a suitable program, an MPU can be used to control or operate almost all conventional industrial processes, this will eventually mean that all servicing engineers involved in electrical or electronic servicing, while not needing knowledge of programming techniques, will benefit from an understanding of the basic operation of the MPU.

18   Logic-system problems

Introduction

This chapter aims to introduce the reader to some of the common faults which may affect modern digital systems and to give an overall view of the methods of approach to problems which may be suitable.

To fault find ('debug') successfully and quickly there are no short cuts. If the problem persists after faults which have occurred before have been eliminated, there is *no substitute* for a *thorough understanding* of what the circuit is supposed to do. Test the circuit by placing it in the mode where it makes errors most frequently. Compare how it works with the way it was designed to work.

Remove faults as soon as they are detected, one at a time.

In combinational circuits, errors usually occur for particular values of input variables. Set the circuit up using these input variables and then investigate step by step until a gate is found which is not working as specified.

Sequential circuits are more complex and consequently pose more problems for the servicing technician. It is important to remember that most sequential circuits are controlled by a 'master-clock' timing sequence. Some errors occur only when certain flip-flops are in 'set' or 'clear'. The steps involved in debugging are

 i) understand the circuit;
 ii) set up the circuit to give fault conditions;
iii) expect multiple errors;
iv) compare the actual operation with the theoretical.

When testing equipment containing integrated circuits, take care not to short circuit pins by using large test probes. Avoid the use of excessive heat when soldering, and *always* switch off the power supply before removing or replacing an integrated circuit, otherwise excessive surge currents, which can destroy the integrated circuits, may occur. Remember also that most power supplies contain large smoothing capacitors which can hold the supply voltage at a reasonable level for up to hours after switch-off, so it can be good practice to ensure that these are discharged through a large-value resistor.

For fault finding on integrated circuits, use a logical method each time, such as

 i) check the power supply at the *integrated-circuit pins*;

ii) make sure that the required input is present at the integrated-circuit pin indicated on the diagram;

iii) check for a suitable output;

iv) check visually and with a meter for any open or short circuits in the track to the integrated circuit.

There are a number of specialist servicing tools available for use with integrated circuits, such as integrated-circuit inserters, test clips (to prevent accidental shorting of pins), logic probes, logic-signal injectors, etc. These should be used whenever possible.

When removing an integrated circuit by desoldering, *always* use a desoldering tool to remove solder from each pin in turn until the integrated circuit can be lifted out.

Remember: 'if in doubt, don't' – it is possible for *all* the integrated circuits in a system to be destroyed by an inadvertent error or an unskilled operator.

Power-supply faults

Whatever the system problem, it is advisable to check the power supply first as a matter of routine.

Voltage high or low

The output voltage of a power supply is easily checked using a multimeter. The readings can then be checked against the system specification.

If a fault is detected it is generally due to a breakdown in the power supply, but not always. Disconnecting the power supply from the system quickly determines whether the fault is internal or external to the power supply.

Faults in the power supply are generally due to the failure of the power-regulating transistor(s) – an open-circuit regulating transistor giving low output and a short-circuit transistor a high output. Having rectified the fault ensure that, if it was the logic system which caused the fault, this is also rectified before reconnecting the power supply.

Output noise

The symptoms of electrical noise in a logic system are generally intermittent faults. The problem in logic occurs with the bistables, since a short pulse appearing on the input to a bistable can cause it to be set or reset. The nature of noise is its variability, so in practice bistables are only affected by the noise occasionally, since most of the pulses will be too small to cause a change.

Noise on supplies should be measured using an oscilloscope and be checked against the system specification. Care must be taken when making these measurements, as the noise may include large voltage spikes of very short duration. It is these which are most likely to cause system malfunctions. The oscilloscope should be used at a high brightness level and with both high and low *Y*-amplifier settings. Even under these conditions the oscilloscope must be watched carefully if the trouble is to be found.

Noise in power supplies often originates with the zener diodes. Usually this is minimised by putting a capacitor across the zener, but if this capacitor fails then the output will be noisy. Note that system-generated noise can make the power supply appear noisy. To minimise this effect, large electrolytic capacitors are often found across the output of power supplies. Being electrolytic, these capacitors are prone to failure.

Noise apparently originating in power supplies can in some cases originate with noise on the incoming mains supply. There are two main precautions taken against this. Firstly, a transformer with an earthed screen between primary and secondary windings prevents direct capacitative coupling. Secondly, transient-suppression devices are often connected across the transformer windings – these may be devices based on zener diodes, voltage-dependent resistors, or just *R–C* networks. These devices are difficult to test, and replacement is the simplest approach.

Logic modules

General method of approach
The method of localising a system fault down to a particular card and then down to a particular module is a difficult one, as it depends on the type of system involved. It has already been stated that a power-supply check for correct voltage levels should be undertaken initially. Consideration should then be given to the machine conditions which can be observed.

The machine operator's description of the machine's behaviour before failure may be useful but should be treated with care, as the information given may be doubtful. Machine conditions at failure coupled with a reasonable overall system knowledge can indicate the system area where failure has occurred.

Having located the system area or card which appears to be faulty, you may then be lucky and find that exchanging for a known good card will rectify the problem; however, if the fault is external to the card, you may damage the spare card too. It is more prudent to check that the output signals from the card which should be present are in fact present. If they are present and correct, the fault is external to the card. If they are not present or are at incorrect levels, the fault could be either internal or external to the card and more investigation is required. Eventually, the problem will be reduced to a particular device or logic module which is faulty and can be replaced.

Types of fault
Logic modules such as TTL or integrated circuits generally fail in a catastrophic fashion – i.e. they will not work at all. These are fairly easy to find and are obviously faulty when located. Less often, modules can

exhibit 'soft' failures, i.e. the modules work partially. These are more difficult to locate and require closer investigation to prove that they are faulty.

The usual type of soft failure is to give output logic levels which are outside the specified logic-level bands. In some cases these can give solid faults, but they can also give intermittent malfunctions. Another type of soft failure can produce the type of output shown in fig. 18.1. This can be especially noticeable when the gate input has a long rise or fall time. The extra pulse on the faulty output can give trouble on the trigger inputs of flip-flops.

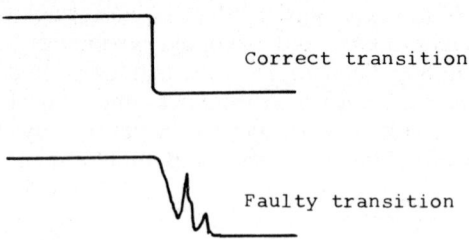

Fig. 18.1 Faulty transition

Methods of diagnosis

One of the most important diagnostic tools for logic is a logic probe. This is a simple device which allows rapid checking of logic levels. A number of different types of logic probe are made by many manufacturers, and two representative types are described below.

Type 1 Designed to check that logic levels conform to specification. A lamp is illuminated when the input signal is at a 1 level and a second lamp (generally of a different colour) is illuminated for an 0 level. No lamp is illuminated for a signal which is outside the specified band. Some probes allow the band limits to be pre-set.

Type 2 There are no clearly defined logic-level limits, but a logic 1 illuminates the lamp and a 0 extinguishes the lamp.

If the probe input is 0 and a short-duration pulse going to logic 1 and returning to 0 appears, the probe lamp will be lit for a short period of time to indicate the presence of the pulse. This type of probe can readily indicate the presence of very short pulses, since the lamp is lit for a minimum period of time defined by a monostable. Similarly, a pulse going to 0 will extinguish the lamp for a minimum period of time.

In selecting a probe, it must be ensured that it is suitable for use on the logic levels found in the system to be checked. Most probes are suitable for use on nominal logic levels of 0 V and 5 V but not for CMOS levels. Many logic probes are built by the people who are going to use them, which need not be a difficult exercise, particularly if the probe is constructed using the types of module which it is desired to test.

Multimeters and oscilloscopes can also be used for diagnosis but are generally less convenient than logic probes. Oscilloscopes are vital for the more difficult faults, as will be described later in the chapter.

One useful technique in testing individual gates is to force gate inputs to 0 V by putting a short circuit to earth on these inputs. This must be attempted only where it is *known that damage will not be caused*. Whether damage will ensue depends on the type of gate, bistable, etc. which is driving that gate input. Two types of output circuit are given in fig. 18.2.

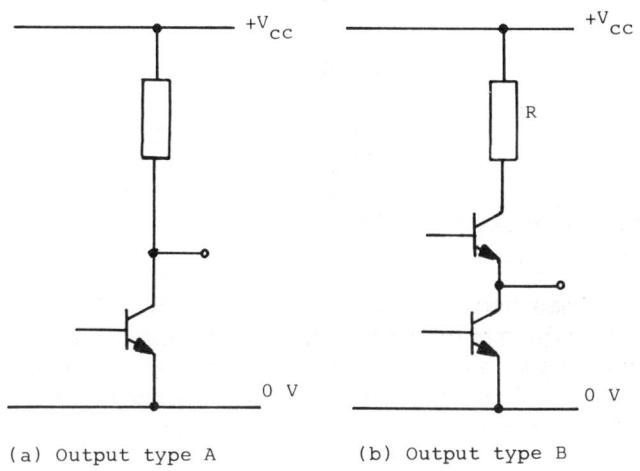

(a) Output type A   (b) Output type B

Fig. 18.2 Types of output circuit

Clearly the type-A circuit can have its output shorted to 0 V without damage, since the current is limited by the collector resistor. Output type B, on the other hand, is intended to provide large output currents and the resistor R is not always large enough to limit the current through the upper transistor when the output is short-circuited to 0 V – the resistor is present for a different reason.

Using this technique, it is possible to work through the truth table for a module without any need for desoldering. There are certain probes which can inject either 0 or 1 without any damage.

System noise

Due to power supply
This has already been covered.

Decoupling capacitors

Some types of circuit and some types of logic modules create noise on the power-supply rails. If the level of this noise is great enough, bistables may change state with consequent malfunctions.

Decoupling capacitors are connected between the power-supply rails adjacent to noise-producing modules, to reduce the amount of noise caused. Occasionally, these capacitors give trouble which can be difficult to locate. The simplest test method is probably to connect a known good capacitor in parallel with each decoupling capacitor in turn until the noise is reduced to acceptable limits.

Earthing faults

These are generally difficult to find. Problems caused by earthing faults are generally intermittent. Investigation is best conducted using an oscilloscope earthed to a known good mains earth and measuring the noise at various earthing points throughout the system. A very noisy measurement indicates a lack of proper earthing.

When the earth fault is located, ensure that the repair replaces the earthing as it was originally. It is unwise to add a new earth wire in a new position, because of the possibility of setting up 'earth loops'. An earth loop can apparently cure problems but will almost certainly create others.

A good earthing system for an equipment located in several cabinets uses only one earthing point to mains earth and only one wire from each cabinet to that earth point. There are thus no loops in the system. This is illustrated in fig. 18.3, where a wire which would cause an earth loop is shown dotted.

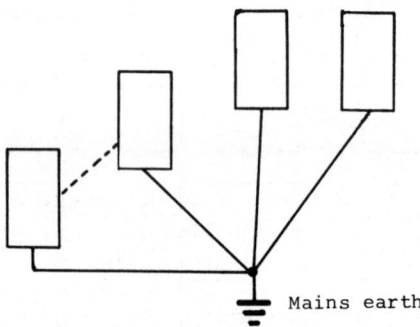

Fig. 18.3 Earthing

Induced noise

Noise can be induced into a wire by another wire running close to it which carries rapid changes of voltage or current.

This is not normally a problem which occurs in an established system, but it can occur as a result of modifications to that system or a nearby system where new cables have been laid near to the system's signal cables.

124

Glitches

The term 'glitch' is applied to a small unwanted pulse appearing in a logic signal. Glitches often occur regularly but their size varies, so occasionally they are large enough to trigger bistables etc., causing intermittent faults.

Once their presence is suspected, glitches can be located using an oscilloscope. This can involve extremely careful work with the oscilloscope, using a high level of brightness and a high time-base speed, as a glitch may be present for a very short time indeed. An oscilloscope with a delayed time-base is virtually essential.

Glitches occur for various reasons, as indicated below.

Design error

Figure 18.4 illustrates a common form of design error which will create a glitch. It is caused by the time difference between the rising edge of $\bar{A}$ and the falling edge of B. Since the falling edge of A (rising edge of $\bar{A}$) causes B to change, any delay in B creates the situation shown on an expanded time scale in fig. 18.5.

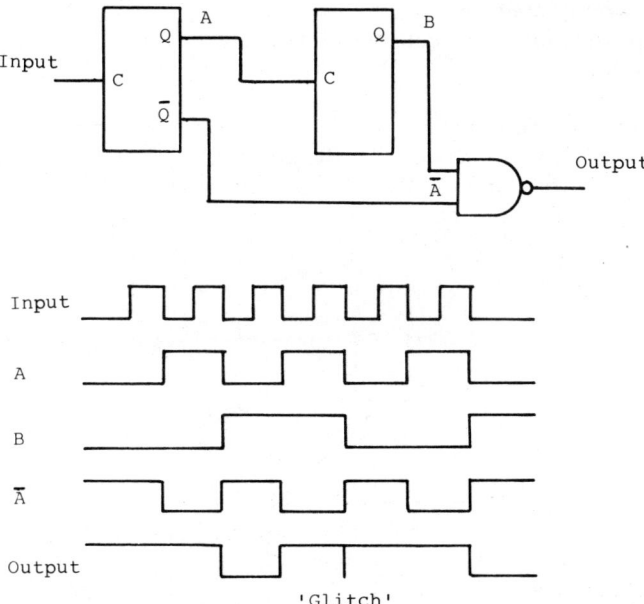

Fig. 18.4 Glitch generation

Race conditions

Occasionally two signals are gated together which can produce glitches in the output if the actual timing of the signals varies slightly. The term 'race

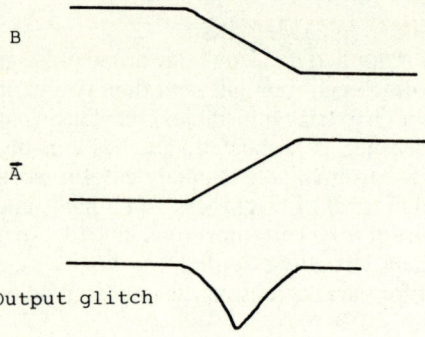

B

$\overline{A}$

Output glitch

Fig. 18.5 Glitch

condition' indicates that whether a glitch is produced or not depends on which signal arrives first.

Soft failures of modules
This type of failure can create one or more glitches, as indicated earlier in this chapter.

Switch or relay contacts
When a pair of contacts closes, the contacts tend to bounce slightly, creating a series of pulses. This can also occur as the contacts break, but generally to a lesser extent.

19   Power supplies

Introduction

There are four main types of power supply in use:

  i) a.c. input, d.c. output;
 ii) d.c. input, d.c. output;
iii) a.c. input, a.c. output;
iv) d.c. input, a.c. output.

Most industrial electronic equipment in use at present contains devices requiring a d.c. power supply, and type (i) is the most likely.

All power supplies can be represented by the block-diagram layout in fig. 19.1.

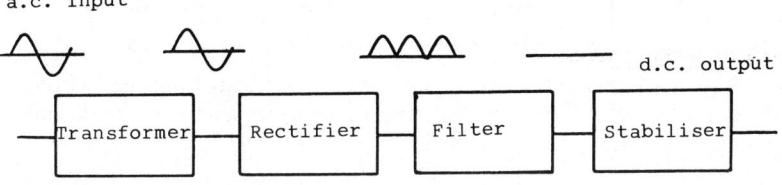

Fig. 19.1 Power-supply block diagram

Basic rectification

An unregulated power supply consists of a transformer, a rectifier, and a smoothing (filter) circuit. This is shown in fig. 19.2.

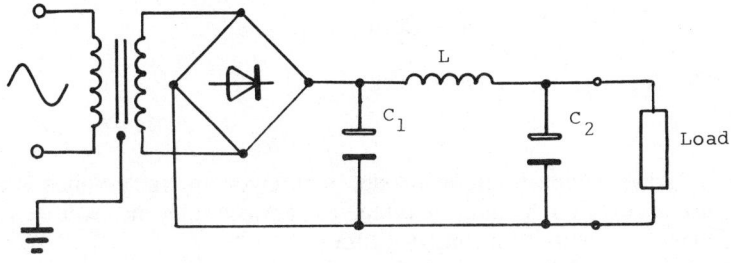

Fig. 19.2 Basic unregulated power supply

For most applications this simple circuit is not good enough, since the d.c. output has a ripple superimposed on it, and the output varies with load current and temperature.

To overcome the problem of variation of output voltage, a stabilised power supply or electronic regulator, using feedback, is used in conjunction with the unregulated power supply.

Stabilising methods

There are two main methods of incorporating a stabilising circuit into a power supply:

a) parallel or shunt stabilisation;
b) series stabilisation.

These basic circuits are shown in fig. 19.3.

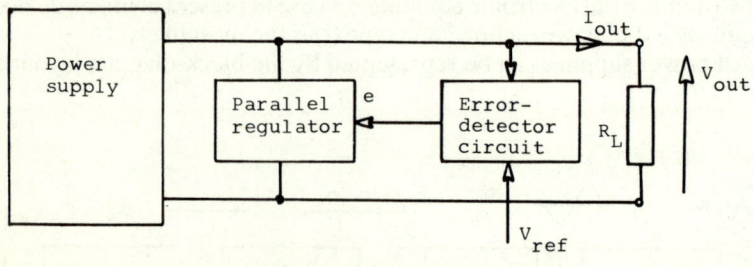

(a) Parallel regulator

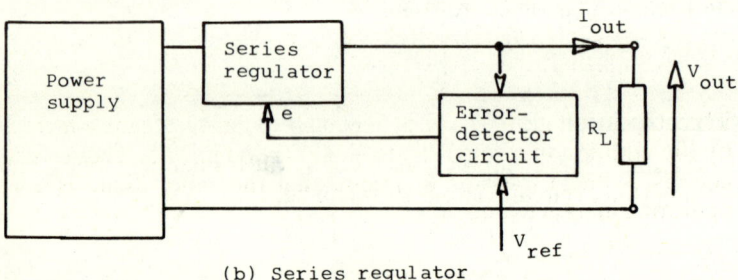

(b) Series regulator

Fig. 19.3 Basic methods of stabilisation

The 'error-detector' circuit produces an error voltage e which is the difference between the output voltage V_{out} and the reference voltage V_{ref}. This error is then used to control a circuit.

In case (a), the error voltage is used to bypass any excess current to ground and keep I_{out} constant, thus $V_{out} = I_{out}R_L$ remains constant. In

128

case (b), the error voltage is used directly to control the amount of current into R_L, and hence V_{out}.

In practice, (b) is preferred since it simply cuts down current from the supply, whereas (a) simply bypasses any excess to ground and the power supply *always* provides maximum output so the parallel regulator must be able to dissipate maximum power when $I_{out} = 0$.

Practical voltage regulators

These take a varying input voltage and produce a fixed regulated output voltage. They are available in commercial packages with a choice of voltage and current ranges, with positive or negative outputs.

Some basic circuit arrangements are shown in figs 19.4 to 19.7.

a) The basic circuit arrangement is shown in fig. 19.4. C_1 and C_2 are decoupling capacitors. (Decoupling or 'bypass' capacitors effectively short circuit unwanted a.c. signals to ground.) C_1 prevents high-frequency instability; and C_2, as well as decoupling the a.c. signal, reduces the output impedance at high frequencies.

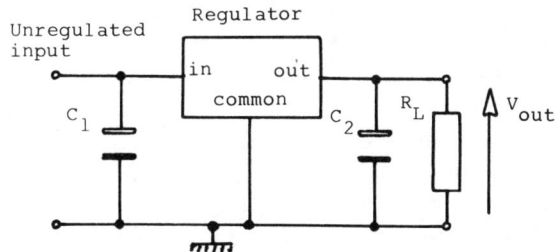

Fig. 19.4 Voltage regulator

b) The output voltage of the basic regulator of fig. 19.4, can be increased by the addition of a zener diode as shown in fig. 19.5. The resistor R acts as a 'bleed' resistor to ensure that the zener diode is always operating in its active region.

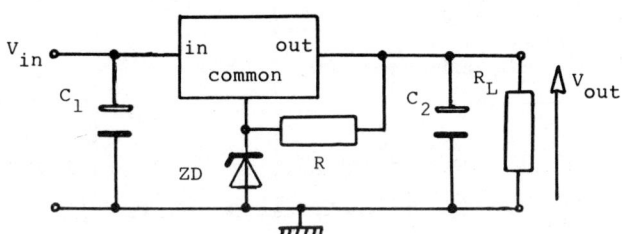

Fig. 19.5 Increasing basic regulator output

129

c) The output current can be increased using a bypass transistor as shown in fig. 19.6.

The value of resistor R should be chosen so that, as I_{in} approaches its maximum value, the voltage drop across R, which is the V_{be} of transistor T_1, exceeds the turn-on voltage, typically 0.5 V for a silicon transistor. Thus the transistor supplies all currents in excess of $I_{in(max)}$, and the output remains regulated.

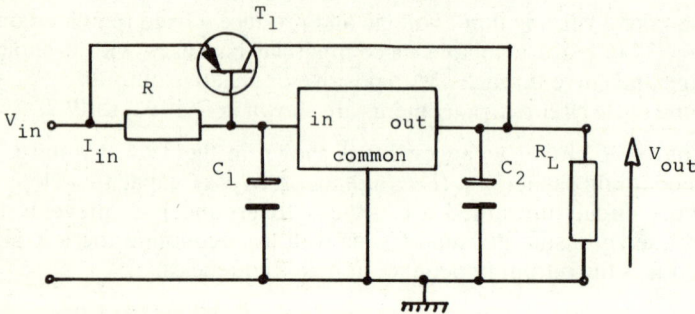

Fig. 19.6 Increasing output current

d) An unregulated voltage which is higher than maximum rated input for the device can be overcome using a simple voltage-regulating zener diode and transistor T_1 as shown in fig. 19.7. The zener breakdown voltage is chosen to be approximately 5 V above the regulated output voltage. Transistor T_1 dissipates most of the power. Transistors such as T_1 are normally power devices and are mounted on heat sinks.

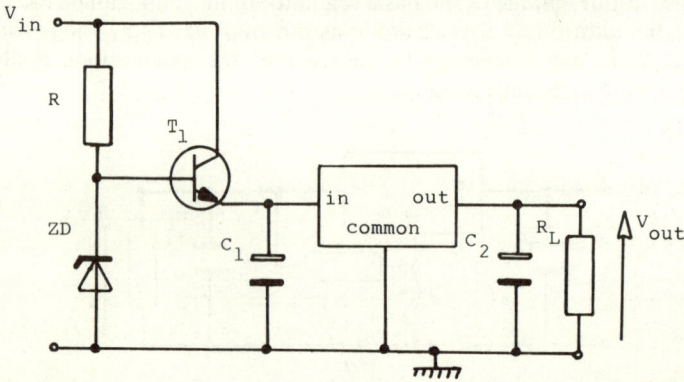

Fig. 19.7 Reducing regulator input voltage

130

Dual power supplies

Many applications in electronics require the use of a dual power supply: a voltage positive with respect to a common and a voltage negative with respect to common. Figure 19.8 shows a typical circuit.

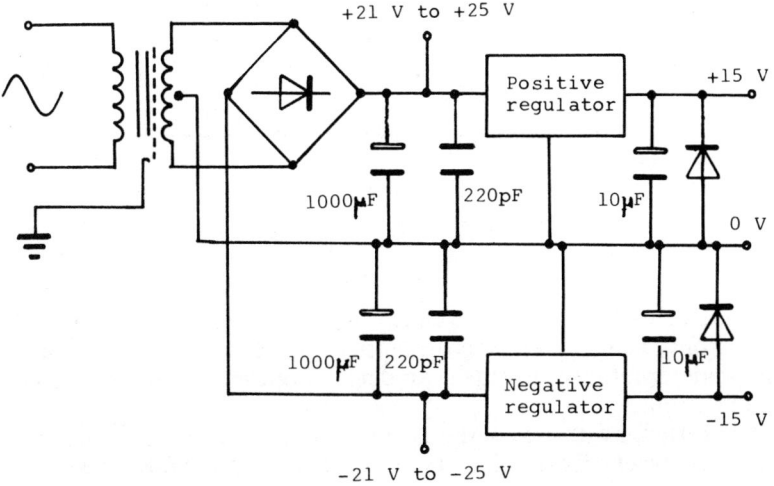

Fig. 19.8 Dual power supplies

Switched-mode power supplies

In the basic voltage regulators shown in figs 19.4 to 19.7, the devices are operated in the linear mode, i.e. somewhere between full on and full off. Since most regulators consist essentially of a transistor acting as a variable resistor, this will dissipate a large amount of power which is proportional to load current and this means that efficiencies greater than 50% are impossible.

Increasing use is now being made of switching regulators which can achieve efficiencies of 75%.

Basic switching regulator

The basic circuit of fig. 19.9, in block-diagram form, has a solid-state switch which switches power either on or off, without consuming energy, depending on the output of the error-detecting circuit.

S is an electronic switch, such as a fast-switching transistor.

Inductor L – known as a 'swinging' inductor – stores energy while S is closed.

With S open, no input current flows. With S closed, the full input voltage is applied across diode D (reverse-biased) and the full voltage

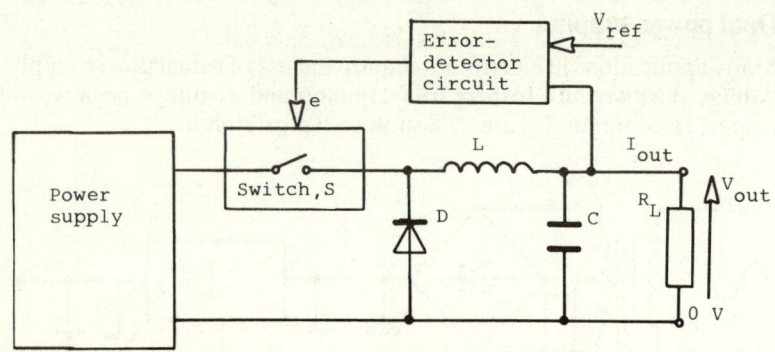

Fig. 19.9 Switching regulator

appears across L and the current through it increases exponentially as capacitor C charges and raises the output voltage. As soon as V_{out} reaches the predetermined value, the error detector opens S and switches off the input current.

The inductor will prevent the current through it dropping to zero, since the magnetic energy stored in it must be dissipated and this will keep the current flowing through L in the same direction. Current now flows in the diode D, continuing to charge C and feed R_L.

When the current supplied by L drops below I_{out}, the capacitor C supplies the load. This lowers the output voltage and eventually the error-detecting circuit switches S on again and the cycle is repeated.

Faults Higher 'ripple', poorer response to surges, noisier (hiss), and r.f. interference due to the switching action. (R.F. or radio-frequency interference is interference carried or radiated by the mains at frequencies referred to as 'radio frequencies' – about 100 kHz to 1 GHz, although the exact limits are arbitrary.)

Appendix 1: revision of basic electrical principles

Prefixes for units

In the field of electrical and electronic engineering, many different units are used. Very often the basic units are either too large or too small for convenient use. Multiples and submultiples of the basic units are then adopted instead.

The following table gives the more common prefixes for units.

Notice also the alternative use of multipliers of powers of ten – this method can be very useful in calculations.

	Prefix	Symbol	Factor
Multiples	giga	G	10^9, one thousand million
	mega	M	10^6, one million
	kilo	k	10^3, one thousand
Submultiples	milli	m	10^{-3}, one thousandth
	micro	μ	10^{-6}, one millionth
	nano	n	10^{-9}, one thousandth millionth
	pico	p	10^{-12}, one millionth millionth

Electrical potential

An essential part of any electric circuit is a source of e.m.f. (electromotive force).

Sources of e.m.f. are any devices which can convert non-electrical energy into electrical energy. The modes of conversion include

a) mechanical–electromagnetic, e.g. the alternator;
b) chemical, e.g. the lead–acid battery;
c) photoelectric, e.g. the solar cell.

If a source of e.m.f. is connected into a closed circuit, an electric current will flow. There is a potential difference (p.d.) between any two points of such a circuit.

The basic unit of e.m.f. and p.d. is the volt (unit symbol V). Their respective quantity symbols are E and V.

The difference between e.m.f. and p.d. can be summarised as follows.

The e.m.f. of a circuit is the total number of volts required to produce a current through the whole circuit, including the source.

The p.d. between two points is that part of the e.m.f. which is required to send the current through the portion of the circuit included between those points.

In heavy electrical engineering, the range of voltages encountered is from volts to megavolts. In electronic engineering, the range of voltages encountered is from volts to microvolts.

Current and charge

In a closed electric circuit, current will flow between any two points which have a p.d. maintained between them.

Conventionally, current flows from the more positive potential point to the less positive potential point.

Electrons carry a negative charge. They move in a circuit from negative to positive.

Positive charge carriers (called 'holes') also exist. They move in a circuit from positive to negative.

Current flow in a conductor can be considered as the movement of electrons in a definite direction along the conductor.

The unit of electrical charge is the coulomb. The unit symbol is C and the quantity symbol is Q (or q).

Experiment shows that the quantity of charge carried by one electron is 1.6×10^{-19} C.

The unit of current is the ampere. The unit symbol is A and the quantity symbol is I.

Although currents of a few tens of amperes (mean) are found in power supplies for large equipments, the normal currents in electronic circuits are measured in milliamperes, and even as low as nanoamperes.

Resistance and resistivity

The ohm, unit symbol Ω, is the resistance (quantity symbol R) between two points of a conductor when a constant voltage of one volt applied between these points produces in the conductor a current of one ampere. Resistance is the opposition to current flow.

The range of values of resistance commonly encountered is from many megohms to a few milliohms.

The resistance of a component is directly proportional to the resistivity (symbol ρ) of the material from which the component is made (this is a constant for that particular material). It is also directly proportional to the length of the component and indirectly proportional to the cross-sectional area. Hence

$$R = \frac{\rho l}{A}$$

134

The unit of resistivity is the ohm metre (Ωm)

From the definition of the ohm, it is clear that, for a complete circuit,

$$\text{total resistance} = \frac{E}{I}$$

from which $E = RI$ and $I = \dfrac{E}{R}$

where R is the total resistance.

Also, for a part of a circuit (that part having resistance equal to R),

$$R = \frac{V}{I}$$

from which $V = IR$ and $I = \dfrac{V}{R}$

where V is the p.d. across that part of the circuit and I is the current flowing into that part of the circuit.

These relationships are commonly known as Ohm's law.

Inductance

Whenever a circuit opposes a change in current (increasing or decreasing), the circuit is said to possess the property of inductance.

When the current tries to increase, the inductance attempts to prevent the increase. When the current tries to decrease, the inductance attempts to prevent the decrease.

An inductor is a component specifically designed to have large inductance. An example is a closely wound coil.

When the current increases in an inductive circuit, the circuit stores energy in a magnetic field. When the current decreases, the circuit returns the energy from the magnetic field.

The unit of inductance is the henry, unit symbol H, and the quantity symbol is L.

In electronics the magnitudes of inductances encountered are henry, millihenry (mH), and microhenry (μH).

Capacitance

Capacitance is the ability of a device to store electrical charge. A device with capacitance also opposes voltage changes in a circuit.

Electrical devices which are used to add capacitance to a circuit are called capacitors.

Almost any arrangement between two conducting bodies that are insulated from each other can come under the heading of 'capacitor'. It is normal, however, to restrict the term to two or more adjacent plates separated by an insulating material which is given the special name 'dielectric'.

The effect of capacitance is present in every electrical circuit whenever the circuit voltage changes.

The basic unit of capacitance is the farad, unit symbol F, and the quantity symbol is C.

The farad is an extremely large quantity, so the units normally used are the microfarad (μF) and the picofarad (pF).

D.C. and a.c. current flow

As already mentioned, current flow in a conductor can be considered as the movement in a definite direction of electrons along the conductor.

It can easily be shown that the number of electrons passing a unit cross-section per second for a conductor carrying 1 A is

$$\frac{1}{1.6 \times 10^{-19}} = 6.25 \times 10^{18} \text{ electrons per second}$$

To understand the essential difference between d.c. current flow and a.c. current flow, consider the following:

a) If a coulomb of electrons moves past a point in a conductor in one second, all of the electrons moving in the same direction, then the current flow is *one ampere d.c.*

b) If, however, a half coulomb of electrons moves in one direction past a point in a *half of a second* and then reverses direction and moves past the same point in the opposite direction during the next half second, then a total of one coulomb of electrons passes the point in one second but this current flow is *one ampere a.c.*

The most conveniently available voltage supply is a.c. In electronics, the voltage required is mostly d.c.

Waveforms

Waveforms are pictures of voltage or current variations with time.

D.C. waveforms are straight lines, since neither the voltage nor the current varies for a given circuit (see fig. A1.1).

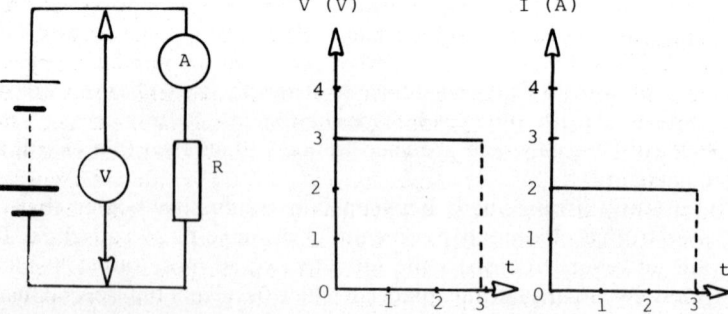

Fig. A1.1 D.C. waveforms

136

To demonstrate a.c. waveforms, consider the circuit in fig. A1.1 to be modified so that the meters are centre-zero type and a four-pole switch is included so that the battery connections can be reversed at regular intervals. If waveforms are drawn for each position of the switch for the relevant period, they will consist of straight lines – above and below zero. By connecting the ends of such lines to form continuous lines, waveforms of voltage and current will result.

These waveforms show that the current and voltage are a.c. rather than d.c., since they indicate the changing direction of current flow and the reversal in polarity of the voltage (see fig. A1.2).

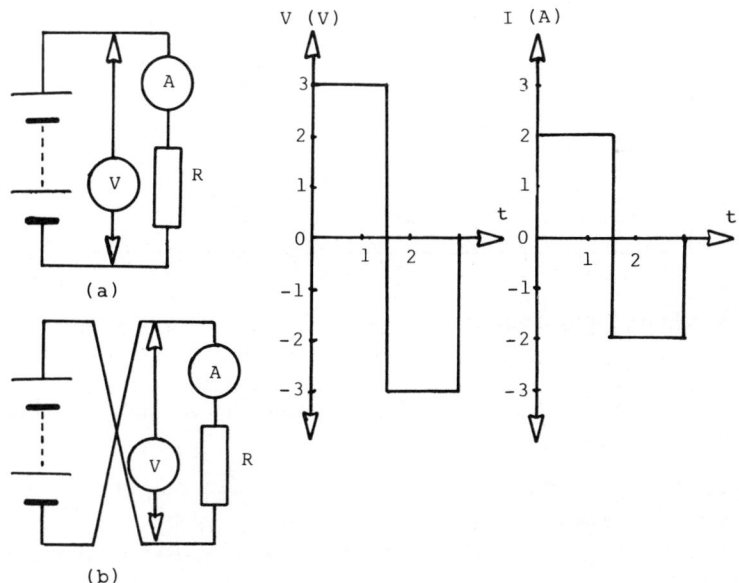

Fig. A1.2 A.C. waveforms

The waveforms in fig. A1.2, consisting of regular straight lines, are in fact square waves. These waveforms (or the closest approximation to them) appear extensively in electronics – especially in logic circuits.

In most cases waveforms are curved, representing gradual changes in voltage and current. The waveforms of the a.c. mains supply have a curved form, representing gradual changes in voltage and current – first increasing then decreasing in value for each direction of current flow. These particular waveform shapes are represented by sinusoidal curves.

The sinusoidal curve, or sine wave, is the basic waveform met with in electronics. Other types of waveform encountered are square waveforms, sawtooth waveforms, and pulsed waveforms (see fig. A1.3).

The waveforms drawn in fig. A1.3 are shown as being symmetrical about the zero-voltage axis. This is not always the case. In many cases a.c.

137

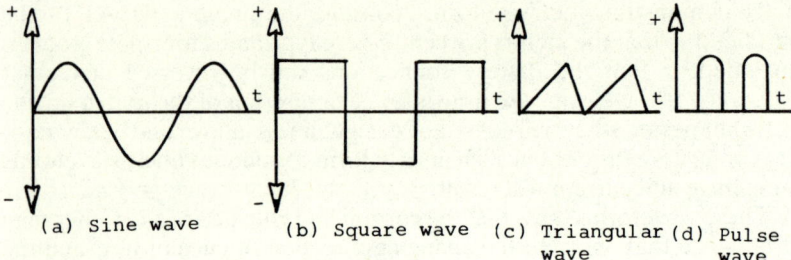

(a) Sine wave   (b) Square wave   (c) Triangular wave   (d) Pulse wave

Fig. A1.3 Various waveforms

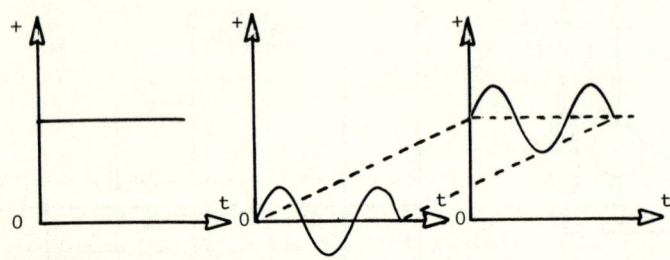

Fig. A1.4 Superimposed waveforms

and d.c. are added together, with the result that the a.c. axis shifts, resulting in 'superimposed a.c.' (see fig. A1.4).

Waveform parameters

Consider the sinusoidal waveform as drawn in fig. A1.5.

A.C. current first rises to a maximum and falls to zero in one direction, then rises to a maximum and falls to zero in the opposite direction (see fig. A1.5(a)). This completes a 'cycle' of a.c. current, and the cycle is repeated as long as the current flows.

Similarly, a.c. voltage first rises to a maximum and falls to zero in, say, the positive direction, then rises to a maximum and falls to zero in the negative direction, to complete one cycle.

A waveform that completes one full cycle is said to have an angular displacement of 360 degrees (360°) from a starting point of zero degrees (0°). (This notation is developed from the concept of sinusoidal waveforms and the method of producing a.c. voltages via rotating machines. A study of this aspect is outside the scope of these notes.) Instead of degrees, radian measure may be used:

$$1 \text{ radian} = \frac{360°}{2\pi} = 57.3°$$

i.e. there are 2π radians in one revolution.

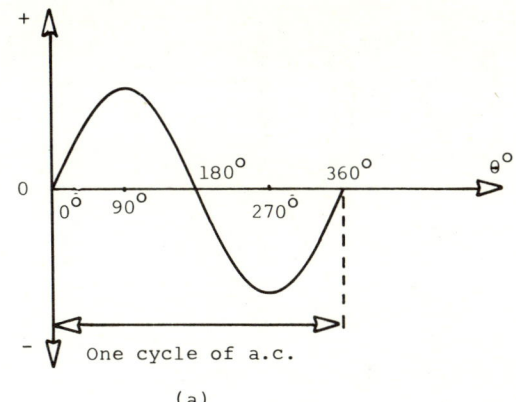

One cycle of a.c.

(a)

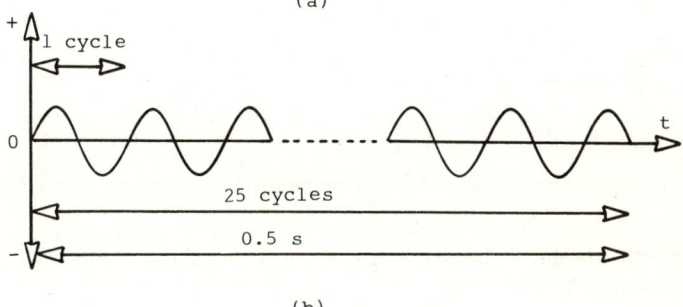

(b)

Fig. A1.5 Sinusoidal waveforms

The 'frequency' of a waveform is the number of cycles completed in one second.

The unit of frequency is the hertz, unit symbol Hz. Note that 'hertz' means 'cycle per second'. The quantity symbol for frequency is f.

Figure A1.5(b) shows 25 cycles of sinusoidal waveform completed in ½ second. The frequency of this sinusoidal waveform is said to be 50 Hz, i.e. the waveform completes 50 cycles in one second. This is the frequency of the a.c. mains supply.

In electronics, the useful frequency range is from zero frequency (d.c.) up to and beyond many millions of hertz.

The 'periodic time' of a waveform, quantity symbol T, is the time the waveform takes to complete one cycle. Thus, for the a.c. mains, where $f = 50$ Hz, the periodic time is

$$T = \frac{1}{f} = 0.02\,\text{s} = 20\,\text{ms}$$

Similarly, if $f = 1$ kHz then

$$T = \frac{1}{f} = \frac{1}{1000} = 1\,\text{ms}$$

139

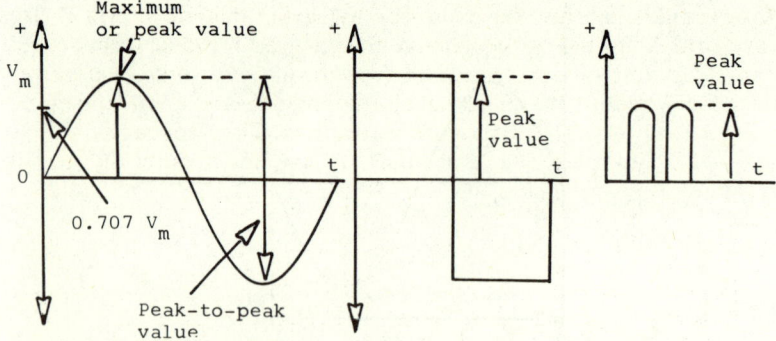

Fig. A1.6 Waveform peak and r.m.s. voltages

The magnitude of waveforms is expressed in different ways, depending on the application (fig. A1.6).

In the case of the mains a.c. supply, the magnitude is expressed as the r.m.s. value (root-mean-square value):

r.m.s. value $= 0.707 \times$ maximum value

The r.m.s. value is used because this is the value of a constant d.c. supply which will dissipate the same power in a given load. Most electrical meters read in r.m.s. values.

Very often, in electronics, the interest lies in the maximum or peak value and so waveforms are often expressed in terms of peak-to-peak values. Such values are easily read from the display on the oscilloscope.

Another property of waveforms that can be very important is the relative phase of one waveform to another. In fig. A1.7, waveform A is said to 'lead' waveform B by an angle ϕ. As can be seen, if the start of the observation is at $t = 0$ then waveform A reaches its first positive peak before waveform B – the angular distance between the waveform is ϕ.

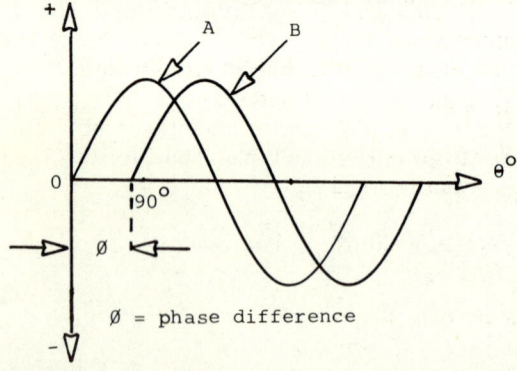

Fig. A1.7 Phase difference

However this can also be expressed by saying that waveform B 'lags' waveform A by angle ϕ (which may be in degree or radian form).

In electronics, non-sinusoidal waveforms are used extensively, particularly repetitive square waveforms and pulsed d.c. waveforms.

Figure A1.8 shows two square waveforms. They appear as a square pulsed d.c. However, as shown earlier, the axis of a.c. can easily be raised by adding appropriate d.c.

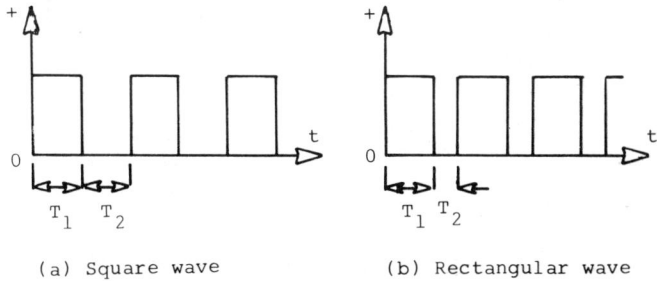

(a) Square wave   (b) Rectangular wave

Fig. A1.8 Square and rectangular waveforms

In fig. A1.8(a), the width of the pulse A equals the width of the relaxation time B before the next pulse arrives. The time (T_1) corresponding to A is termed the 'mark' and the time (T_2) corresponding to B is termed the 'space'.

The ratio of T_1 to T_2 is termed the mark–space ratio. In the case of waveform (a), $T_1 : T_2 = 1$.

Very often a waveform as shown in fig. A1.8(b) is used, in which the mark–space ratio is greater than unity.

Resistance in d.c. and a.c. circuits

Ohm's law applies to a.c. and d.c. circuits containing *only* resistance.

For the circuit in fig. A1.9(a), R_1 is in series with R_2,

thus $R_T = R_1 + R_2$

The current is common to R_1 and R_2,

$\therefore$ $V_{R1} = IR_1$ and $V_{R2} = IR_2$

Then $V = V_{R1} + V_{R2} = IR_1 + IR_2 = I(R_1 + R_2)$

$\qquad\quad = IR_T$

For the circuit in fig. A1.9(b), R_1 is in parallel with R_2,

thus $\dfrac{1}{R_T} = \left(\dfrac{1}{R_1} + \dfrac{1}{R_2} \right) = \dfrac{R_1 + R_2}{R_1 R_2}$

141

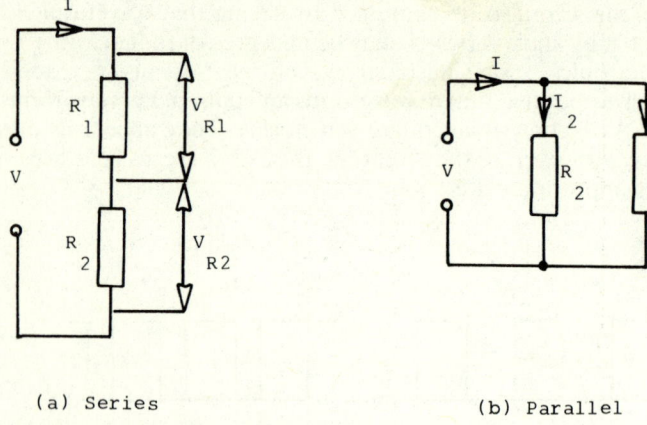

(a) Series   (b) Parallel

Fig. A1.9 Resistor combinations

The circuit current I is split at the junction of R_1 and R_2 so that

$$I = I_1 + I_2$$

The voltage across R_1 is the same as the voltage across R_2 (the voltage is common),

$$\therefore \quad I_1 = \frac{V}{R_1} \quad \text{and} \quad I_2 = \frac{V}{R_2}$$

Then $\quad I = I_1 + I_2 = \frac{V}{R_1} + \frac{V}{R_2} = V\left(\frac{1}{R_1} + \frac{1}{R_2}\right)$

$$= V \times \left(\frac{R_1 + R_2}{R_1 R_2}\right)$$

$$= \frac{V}{R_T}$$

A.C. current and voltage in resistive circuits

In a resistive circuit the current flow follows the voltage exactly: as the voltage increases, the current increases; when the voltage decreases, the current decreases; and at the instant the voltage changes polarity, the current flow reverses its direction. The voltage and current are said to be *in phase*.

In fig. A1.10, the two waveforms have the same frequency and they are in phase, but obviously their amplitudes are not necessarily equal.

Inductance in d.c. and a.c. circuits

Every inductive circuit has resistance, since the wire used in a coil always has some resistance.

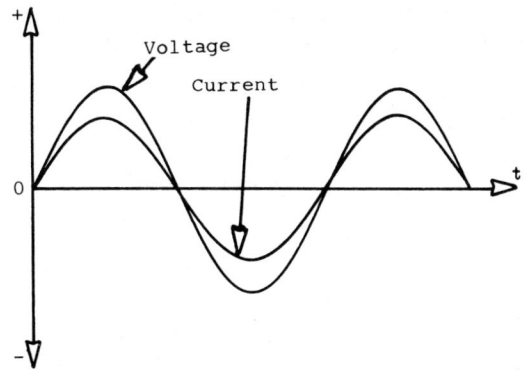

Fig. A1.10 Voltage and current waveforms in resistive circuits

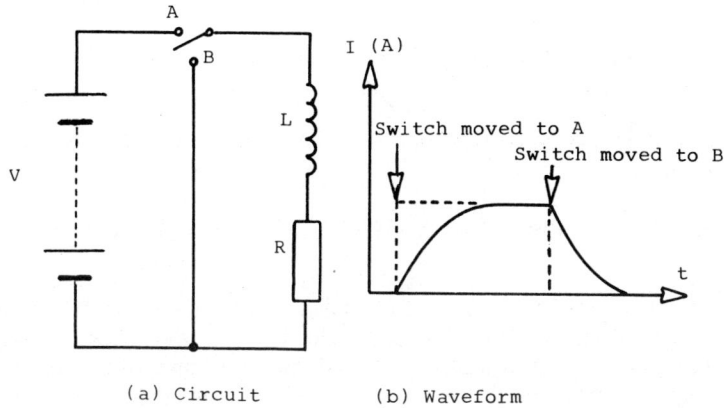

(a) Circuit   (b) Waveform

Fig. A1.11 Rise of current in inductive circuits

Figure A1.11(a) shows an inductor L with its ohmic resistance R in series with it in a d.c. circuit.

When the switch is moved to A, current will begin to flow but the inductor will oppose this flow (since an inductor opposes any change in current); hence the current rise is delayed as shown in fig. A1.11(b). A point will be reached at which full current will flow, $I = V/R$.

When the switch is moved from A to B, the stored electrical energy in the magnetic field about L will be returned to the circuit. In other words, removing the circuit from the battery will be an attempt to reduce the current in the circuit to zero at once. However, the inductor will oppose this change of current by using the energy stored in the magnetic field to induce an e.m.f. which opposes the original change. (This effect is known as Lenz's law.) Thus the current fall is delayed as shown.

Notice that switching on and off the switch from position A might be expected to produce a square waveform, but the inductor distorts this as shown.

The same inductive circuit would affect a.c. voltage and current waveforms as follows (fig. A1.12).

When the a.c. voltage is applied, the current increases as the voltage increases. However, due to the delaying action of the inductor, the current will not reach its maximum d.c. value before the polarity of the voltage is reversed.

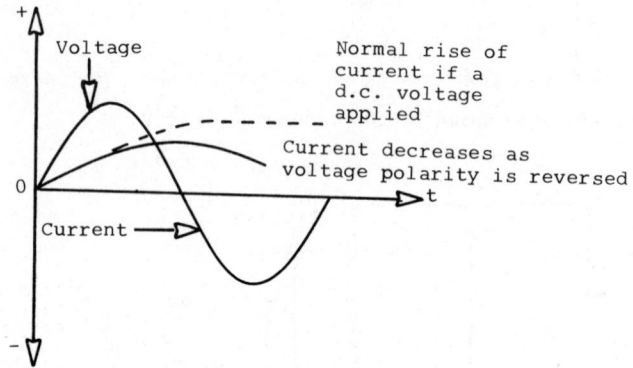

Fig. A1.12 Voltage and current waveforms in inductive circuits

How near to this d.c. value it will reach will depend on the time it takes for the voltage to complete a half cycle. Thus the magnitude of the current depends on the frequency of the voltage: the lower the frequency, the longer the time the current has to increase; hence the larger the current, the lower the frequency. The opposition to current flow in an a.c. inductive circuit is therefore proportional to the inductance *and* the frequency.

This opposition – called the inductive reactance – is expressed by the formula

$$X_L = 2\pi fL$$

where X_L = inductive reactance in ohms

f = frequency in hertz

and L = inductance in henrys

The quantity '2π' is a constant number (it represents one cycle).

As shown, one effect of the inductance is to delay the rise and fall of the current. The current is said to 'lag' behind the voltage.

In an a.c. circuit containing only pure inductance, the current would lag the voltage by 90°. However, all practical circuits contain resistance, so in inductive circuits the current lags the voltage by an angle less than 90°.

The exact amount of lag depends on the ratio of circuit resistance to inductance – the greater the resistance compared to the inductance, the nearer to an in-phase (no phase difference) condition it becomes; the lower the resistance compared to the inductance, the nearer the phase difference gets to 90°.

Inductances in series and parallel

For determining the total inductance in such combinations, the same rules apply as for resistors in series and parallel.

Capacitance in d.c. and a.c. circuits

When a voltage is applied across the terminals of a circuit containing capacitance, the voltage across the capacitance does not instantaneously equal the voltage applied to the terminals.

The voltage across a capacitor is proportional to the charge stored on its plates. On moving the switch in fig. A1.13(a) to A, current will flow into the capacitor. As the charge on the plates increases, the voltage across the plates rises and can be represented by a graph similar to that for the rise of current in an inductive circuit. The voltage is delayed in this case. This is so because a capacitor will oppose a change in *voltage*. When the capacitor is fully charged, current will cease to flow.

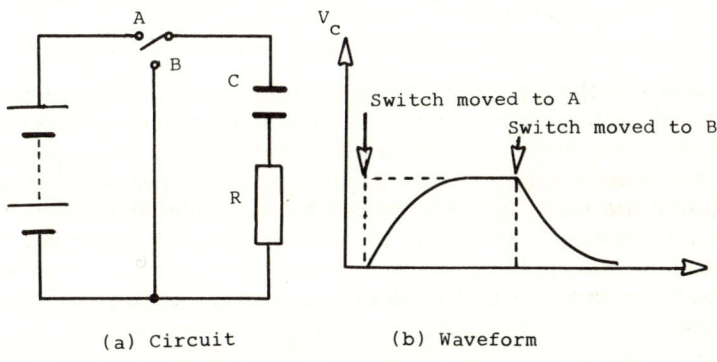

(a) Circuit   (b) Waveform

Fig. A1.13 Rise of voltage in capacitive circuits

Thus, except for the initial charging current, no current will flow through a capacitor in a d.c. circuit; i.e. a capacitor is a d.c. block.

When the capacitor is fully charged, it will hold this charge even if the battery is removed. The charge is said to be stored in the capacitor – this is how a capacitor stores energy.

Moving the switch to B in fig. A1.13(a) will result in this stored charge flowing out of the capacitor. The voltage across the capacitor will fall as the charge is removed.

145

Notice what happens to a square supply waveform when applied to a capacitive circuit having a discharge path available. The waveform is distorted on its leading edge (the 'switch A' section of fig. A1.13(b)) and on its trailing edge (the 'switch B' section of fig. A1.13(b)).

Capacitive reactance is the opposition to current flow offered by the capacitance of a circuit.

The capacitive reactance to d.c. is considered to be infinite.

A.C. continuously varies in value and polarity; therefore the capacitor is continuously charging and discharging, resulting in a continuous current flow in the circuit and a finite value of capacitive reactance.

As a capacitor is being charged, the initial current is high. As the capacitor charges, the capacitor offers more opposition to the charge flow until a point is reached when the capacitor is fully charged and charge flow into the capacitor is reduced to zero (fig. A1.14(a)).

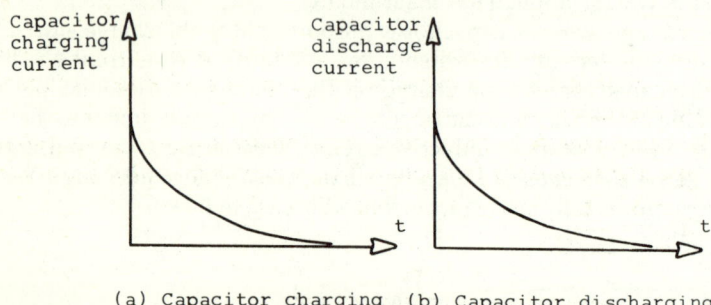

(a) Capacitor charging (b) Capacitor discharging

Fig. A1.14 Variation of current in a capacitor

Similarly, the discharge current is high at the beginning of the discharge, since the voltage of the charged capacitor is high. As the capacitor discharges, however, the voltage available becomes less, resulting in less current flow (fig. A1.14(b)).

Since the charging and discharging currents are highest at the beginning of the charge and discharge of a capacitor, the average current is higher if the polarity is reversed rapidly, keeping the current flowing at high values.

Consider an *R–C* circuit to which a square wave is applied. Figure A1.15 shows the relevant waveforms.

As can be seen, the average current is a function of the level to which the charging current has fallen before the supply voltage changes polarity, charging the capacitor in the opposite direction. Thus, if the polarity switching is very rapid, the average current per half cycle will be high. This corresponds to a high-frequency supply-voltage waveform.

If the frequency is low, average current will be low.

In a similar manner it can be shown that the smaller the value of capacitance, the lower will be the average current.

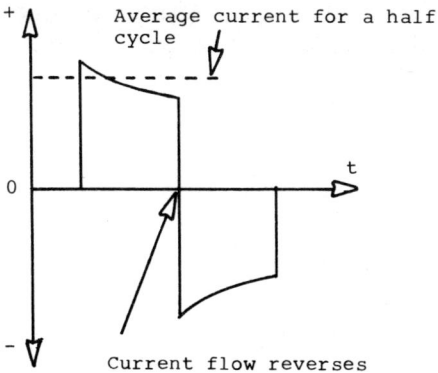

Fig. A1.15 Capacitor current in response to an applied square wave

To summarise: the current flow in a capacitive circuit increases with an increase in frequency or capacitance (assuming all other components remain constant).

The capacitance reactance must therefore be inversely proportional to the capacitance and the frequency. This is expressed by the formula

$$X_C = \frac{1}{2\pi f C}$$

where X_C = capacitive reactance in ohms

f = frequency in hertz

and C = capacitance in farads

In a circuit containing only pure capacitance, the voltage cannot develop until the plates become charged. Thus it is clear that in this case the current *leads* the voltage. In fact, in the case of pure capacitance the current leads the voltage by 90°.

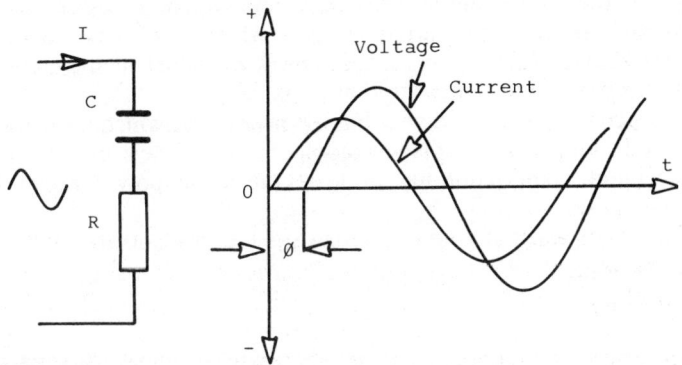

Fig. A1.16 Voltage and current waveforms in capacitive circuits

In more practical circuits where resistance is present, the current leads the voltage by an angle less than 90° (see fig. A1.16).

Capacitors in series and parallel

For capacitors in series (fig. A1.17(a)),

$$\frac{1}{C_T} = \frac{1}{C_1} + \frac{1}{C_2} + \frac{1}{C_3}$$

where C_T = total capacitance

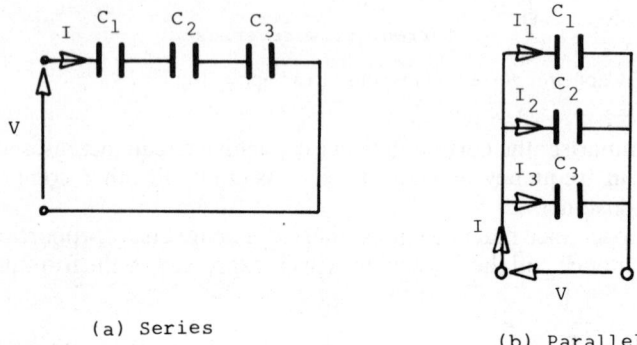

(a) Series

(b) Parallel

Fig. A1.17 Capacitor combinations

For capacitors in parallel (fig. A1.17(b)),

$$C_T = C_1 + C_2 + C_3$$

Power and power rating

When current is flowing in any component which possesses resistance, electrical energy is converted to heat which has to be removed or dissipated to the surrounding space. The rate at which heat is dissipated is known as *power* and is measured in watts.

Every component and device has a power rating which must always be observed; i.e., when replacing components, always ensure that the replacement component has at least the same power rating as the replaced item.

The power dissipated in d.c. circuits is the product of supply voltage and current:

$$P = VI$$

However, in a.c. circuits which contain inductance or capacitance, causing the current and voltage to be out of phase, the power is the product of the current, the voltage, and the cosine of the phase angle (i.e. the angle by which the current leads or lags the voltage):

$$P = VI\cos\phi$$

In heavy electrical applications, the maximum supply current can be reduced without affecting the load power dissipation if the phase angle ϕ can be reduced. As most industrial loads are mainly inductive, ϕ can be reduced by using 'power-factor correction' capacitors. These are suitable value capacitors connected in parallel with the load.

Power loss, or power dissipation, is proportional to the square of the current:

$$P = I^2 R$$

where P is the power loss, I is the current flowing in the component (wire, resistor, etc.), and R is the resistance of the component.

As this power loss usually takes the form of heat generated, it is important in electronics to avoid exceeding the maximum permissible power dissipation of circuits and devices.

Appendix 2: resistors

All conductors have a certain amount of resistance to the flow of an electric current, but in general the term 'resistor' is applied to a device specially chosen for its resistance.

Uses of resistors

Resistors used in electronic equipment are invariably for the purpose of voltage dividing; changing a varying current into a varying voltage, i.e. as a load resistor; or current limiting.

Figure A2.1 shows current symbols for resistors.

Resistors can be divided into two main groups: fixed and variable.

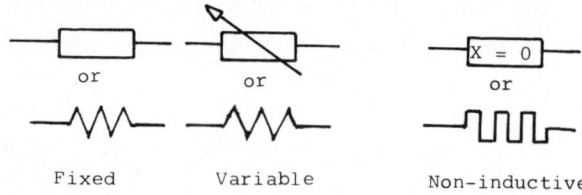

Fig. A2.1 Resistor symbols

Fixed resistors

Carbon-composition resistors
The centre core is a mixture of carbon, binder, and filler in the proportions required to produce the desired final value. The mixture is injected into a ceramic tube and, after the connecting wires have been fitted (fig. A2.2), it is cured in an oven.

A cheaper version does not have the ceramic tube: the carbon compound is coated with a protective lacquer for insulation. Coloured bands are used to denote the resistance value of carbon-composition resistors.

Carbon or metal-film resistors
A film of carbon or metal oxide is deposited on a ceramic former. A helical grove is cut in the film and the cutting is stopped when the required resistance is reached. Connecting wires are fitted and the resistor is coated with protective lacquer.

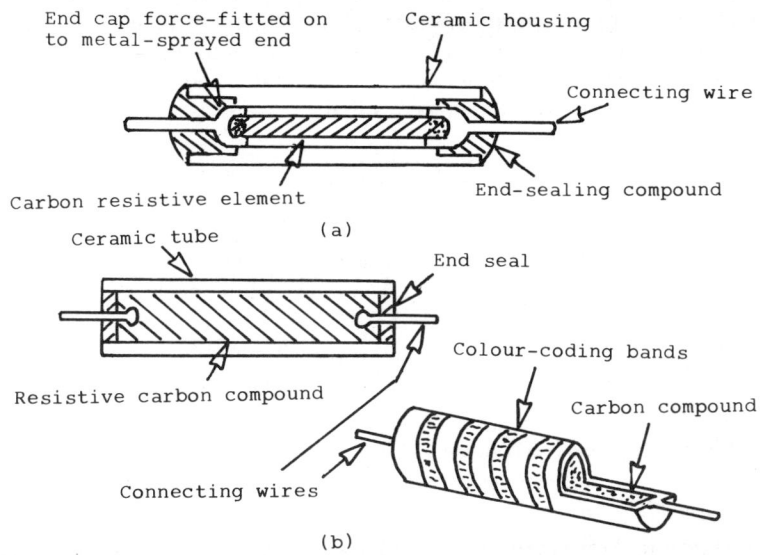

Fig. A2.2 Forms of carbon-composition resistors

Wire-wound resistors

This type consists of wire of a known resistivity wound on a ceramic former and covered by cement or glazed enamel. The resistance value is normally printed on the coating. Wire-wound resistors are used where it is required to dissipate a lot of power, and they usually run hot.

Other types of fixed resistor

Where the disadvantages of instability, noise, and comparatively low power dissipation are troublesome, types of resistor other than the carbon-composition variety must be used.

High-stability resistors   High-stability resistors are generally identified by a salmon-pink band. This band is the fifth band. Stability is not the same as accuracy but is concerned with the change in resistance under working conditions or while being stored on the shelf.

Cracked-carbon resistors   Cracked-carbon resistors are made by coating ceramic rods with conducting carbon. A hydrocarbon, usually methane – previously purified by removing water vapour, oxygen, and carbon monoxide – is passed over ceramic rods at high temperature. Decomposition, or cracking, then occurs at the hot surface of the rods and carbon is deposited.

For the best electrical properties the carbon coating should be thick. High resistances are obtained by spiralling the coating, such spiralling being carried out by machines which automatically produce the desired resistance value. End caps are then force-fitted on to the rod (fig. A2.3) and the carbon film is protected from damage and moisture by the application of numerous coatings of suitable varnishes and paints.

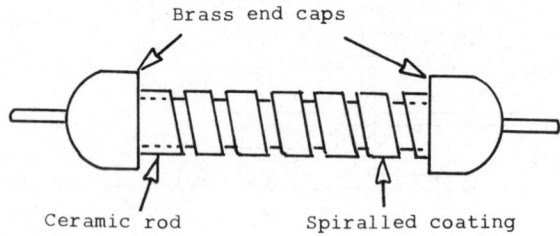

Fig. A2.3 Cracked-carbon resistors

The nominal resistance value of high-stability resistors is often painted on to the resistor body in figures and words, although some manufacturers use the colour-code system. Various power dissipations are available, up to a usual figure of 2 W. The tolerances are usually ±5% or ±10%.

Metal-oxide resistors Metal-oxide film resistors are now displacing cracked-carbon high-stability types because of their increased stability, greater reliability, and smaller size for equal power rating. Since they have been available in quantity, their cost compares favourably with that of other forms of resistor.

The resistor is formed by spraying a solution of stannic chloride and antimony trichloride on to a glass or porcelain rod. On increasing the temperature to red heat, a layer of oxide is formed. By varying the composition of the solution and the thickness of the oxide layer, various resistance values and power dissipations are possible. Spiralling as for the cracked-carbon types is also used to achieve the required resistance.

Colour code for resistors

Resistance values for resistors are indicated by means of a standard colour code (fig. A2.4). Each colour represents a digit as follows:

Black	0	Blue	6
Brown	1	Violet	7
Red	2	Grey	8
Orange	3	White	9
Yellow	4	Gold	±5% tolerance
Green	5	Silver	±10% tolerance

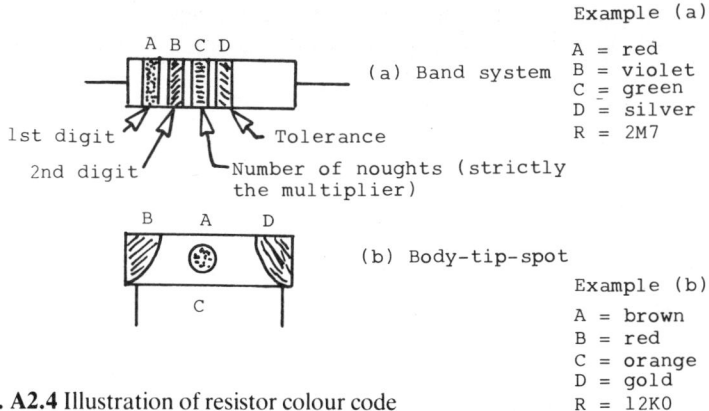

Fig. A2.4 Illustration of resistor colour code

(a) Band system

1st digit
2nd digit
Tolerance
Number of noughts (strictly the multiplier)

(b) Body-tip-spot

```
Example (a)
A = red
B = violet
C = green
D = silver
R = 2M7

Example (b)
A = brown
B = red
C = orange
D = gold
R = 12K0
```

Carbon resistors (insulated)

The band system is used with insulated resistors. Coloured bands are painted on the ceramic tube towards one end. The first digit of the resistance value is indicated by the colour of the band nearest the end (quite often this first band is somewhat wider than the others). The next digit is indicated by the colour of the next band. The colour of the third band indicates the power of the decimal multiplier, i.e. the number of noughts (for example orange represents 10^3, i.e. three noughts). If the third band is gold, the multiplier is 10^{-1}; if it is silver, the multiplier is 10^{-2}. As an example, if the bands are red, red, and gold, the resistance is 2.2 ohms.

If the fourth band is missing, the actual resistance is within ±20% of the nominal value. If the fourth band is silver or gold, then the tolerance is ±10% or ±5% respectively.

Uninsulated resistors

The body–tip–spot system is used with uninsulated resistors. The body colour represents the first digit, the tip colour the second digit, and the spot colour the number of noughts. The tolerance is ±20% unless otherwise indicated.

BS 1852 resistance code

A new British Standard coding is being introduced to indicate the value of many fixed and variable resistors. It gives more information about the resistor, but uses fewer characters.

0.47Ω would be marked R47 100Ω would be marked 100R
1Ω would be marked 1R0 $1k\Omega$ would be marked 1K0
4.7Ω would be marked 4R7 $10 k\Omega$ would be marked 10K
47Ω would be marked 47R $10 M\Omega$ would be marked 10M

and so on.

After this code is added a letter to indicate tolerance:

F = ± 1%   J = ± 5%   M = ± 20%

G = ± 2%   K = ± 10%

Thus

R33M = 0.33Ω ± 20%   6K8K = 6.8 kΩ ± 10%

4R7K = 4.7Ω ± 10%   6K8M = 6.8 kΩ ± 20%

6K8F = 6.8Ωk ± 1%   68KK = 68 kΩ ± 10%

6K8G = 6.8Ωk ± 2%   4M7M = 4.7 MΩ ± 20%

6K8J = 6.8Ωk ± 5%

Power rating

Before a resistor can be chosen for a particular application, the user needs to know more than merely the ohmic value of the resistance. The power rating, i.e. the power-handling capacity, must be known. This is obtained from a knowledge of the maximum current I_{max} that the resistor will be required to pass or, alternatively, the maximum voltage V_{max} likely to be applied. The power rating of the resistor must then exceed $I^2_{max}R$ or V^2_{max}/R to allow a margin of safety.

Voltage rating

Voltage rating is not quite so important as power rating, but there may be instances where it is necessary to consider the voltage rating of a particular resistor, such as in radio circuits involving series-resonant circuits and in voltage-multiplier circuits.

Preferred values

Resistors are available in what appear at first glance to be odd values. This is due to the use of a 'preferred-value' system. The preferred-value ranges ensure that a minimum number of values are required to cover a given range and that the tolerance bands do not overlap each other, i.e. that a nominal resistor of 1 kΩ at its highest value does not exceed the 1.2 kΩ at its lowest value (for 20% tolerance). The old and new systems for a tolerance of ±20% are shown in fig. A2.5.

The range of preferred values of resistors is shown in Table A2.1. Larger values are obtained by multiplying the values given in the table by an appropriate multiple of 10.

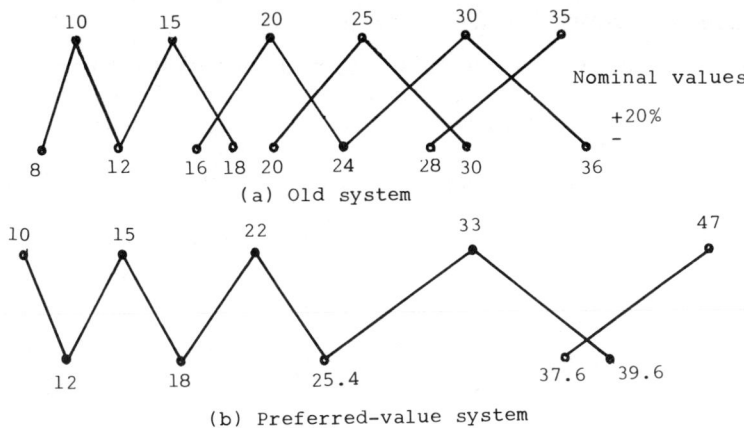

Fig. A2.5 Old and new resistor tolerance systems

Table A2.1 Preferred values of resistors

Tolerance		
20%	10% (silver)	5% (gold)
10	10	10
		11
	12	12
		13
15	15	15
		16
	18	18
		20
22	22	22
		24
	27	27
		30
33	33	33
		36
	39	39
		43
47	47	47
		51
	56	56
		62
68	68	68
		75
	82	82
		91
100	100	100

Variable resistors

There are many instances where it is necessary to vary the resistance of a circuit, e.g. the volume control in radio circuits. Variable resistors generally consist of an adjustable wiper arm making contact with a carbon or wire-wound track.

If two electrical connections are made – one to the wiper arm and the other to one end of the track – a 'rheostat' is formed. This is just a variable resistor. If, however, both ends of the track and the wiper have electrical connections, the device is called a 'potentiometer'. A potentiometer can also be used as a rheostat, and most variable resistors tend to be of this form.

Where variation of the wiper position is controlled by use of a screw and is not normally variable, the resistor is referred to as a 'pre-set' resistor.

Appendix 3: capacitors

A capacitor consists of two conducting plates separated by an insulating material called a dielectric. The capacitance depends on the dielectric, the distance between the plates, and the area of overlap of the plates.

Capacitance is measured in farads (unit symbol F), and the quantity symbol used is C. As the farad is a very large quantity, it is normal to use the microfarad (μF) 10^{-6} F or picofarad (pF) 10^{-12} F. In the case of larger industrial capacitors, the capacitance is rated in kilovolt amperes (kVAr).

Figure A3.1 shows the symbols for capacitors.

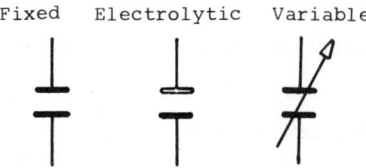

Fig. A3.1 Capacitor symbols

Formulae

$$Q = CV$$

where Q = charge in coulombs

   C = capacity in farads

and V = voltage in volts

$$C = \frac{\epsilon_0 \epsilon_r A}{d}$$

where ϵ_0 = permittivity of free space

   ϵ_r = relative permittivity of the dielectric material

   A = area of overlap (m^2)

and d = distance between plates (m)

The absolute permittivity of a dielectric material is a measure of the ability of the material to store charge when used in a capacitor. Its units are farads per metre (F/m).

The permittivity ϵ ('epsilon') of a material is usually expressed as a multiple of the absolute permittivity of a vacuum or free space, ϵ_0, the multiplying factor being called the relative permittivity of the material, ϵ_r.

i.e. $\epsilon = \epsilon_0 \epsilon_r$

Relative permittivity has no units.

It can be seen that to obtain the largest capacitance in the smallest volume, ϵ_r must be high and d small. The minimum distance between the plates is limited by the applied p.d. – i.e. the V/m will reach a point where the insulating dielectric breaks down and 'flashover' occurs.

$$\text{Capacitive reactance } X_C = \frac{1}{2\pi f C}$$

Some more common methods of colour coding capacitors are shown in figs A3.6 and A3.7.

Capacitor construction

Some of the more common types of capacitor construction are shown in figs A3.2 to A3.5.

Impregnated-paper-dielectric capacitors (fig. A3.2)
The plates consist of two long strips of aluminium foil, separated by strips of insulating paper. Two similar paper strips are placed on the outside of the aluminium strips. The foil-and-paper sandwich is rolled up tightly, vacuum treated, and then impregnated with petroleum jelly or chlorinated naphthalene to increase the breakdown voltage of the dielectric and to exclude moisture.

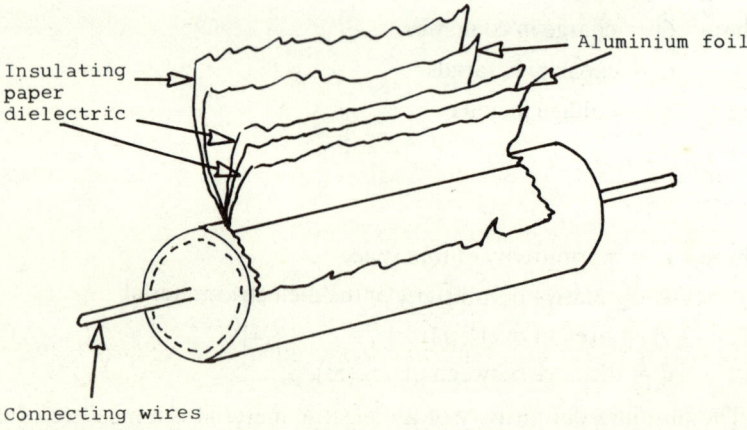

Fig. A3.2 Paper-capacitor construction

The connections to the plates (foil) are made by tinned copper strips. The whole construction is enclosed in a metal case, one end being an insulator through which the connection taps are brought.

Typical capacitance range: $1\mu F$ to $4\mu F$.

Metallised-paper capacitors

The construction is similar to the paper capacitor above, except that the plates are formed by chemically depositing a thin layer of aluminium on the paper. This method of construction allows the capacitor to have a physically smaller size.

Paper-dielectric capacitors are reliable and relatively cheap to produce, but they have high electrical losses and low voltage ratings.

Mica capacitors

Stacked mica The construction of stacked-mica capacitors is shown in fig. A3.3.

Typical capacitance range: 2.2 pF to 10000 pF.

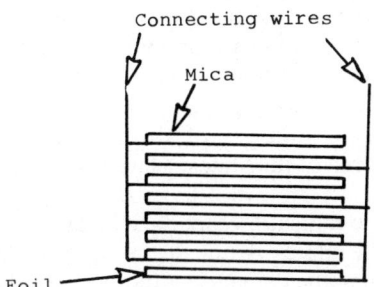

Fig. A3.3 Mica-capacitor construction

Silvered mica This method of construction gives a physically smaller capacitor. Mica sheets are cut to size and silver-oxide powder is coated on to the surface. The sheets are heated and silver forms on them.

Connection strips are fitted after stacking, and the whole assembly is coated in wax.

Typical capacitance range: 10pF to 1100pF.

Mica trimmers The construction of mica trimmers is shown in fig. A3.4.

Typical capacitance range: 2pF to 40pF (single-plate types)
up to 2000pF (multi-plate types)

Electrolytic capacitors

The electrodes consist of two strips of aluminium foil about 0.05mm thick. The foils are separated by two layers of porous paper soaked with

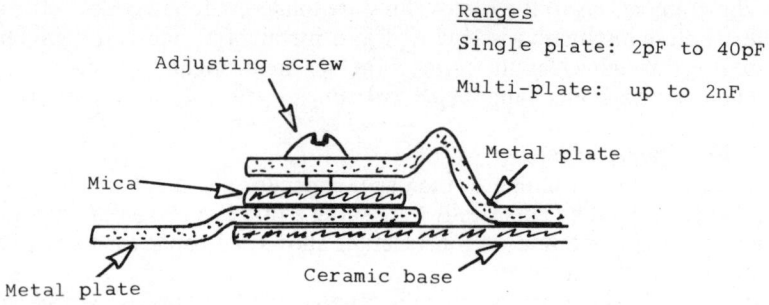

Fig. A3.4 Trimmer-capacitor construction

an electrolyte consisting of a paste of glycol and ammonium tetraborate. The paper also serves to separate the two foils. The assembly is rolled up and the ends are closed with wax. It is then sealed in an aluminium or waxed-cardboard container and connection leads to the plates are taken externally.

A d.c. supply is then connected to the plates and, as a result of electrolytic action, oxygen is produced at the positive electrode (the anode), forming a very thin film of aluminium oxide which, being an insulator, forms the dielectric. The negative electrode (the cathode) has no oxide layer.

Because of the extreme thinness of this dielectric and the comparatively large surface area of the plates, a large capacitance is obtained in a small physical space.

When using electrolytic capacitors care must be taken that the polarity of voltage applied to the plates is not in the reverse direction to the polarity used when forming the dielectric, otherwise the dielectric will be destroyed. To prevent this occurrence, all electrolytic capacitors have + and − clearly marked on their connection tags. Thus these components are said to be polarised.

Typical capacitance range: $1\,\mu F$ to $10000\,\mu F$.

Tantalum electrolytics
These are smaller than the normal electrolytics and can withstand higher voltages. However, they are more costly. They are of similar construction but the plates are of tantalum foil and the electrolyte is sulphuric acid (fig. A3.5).

Air-spaced capacitors
These are variable capacitors. Their capacitance depends on the area of overlap of the plates. One of the plates is fixed and the other is free to rotate. In order to keep the size of the capacitor to practical dimensions, each plate is, in fact, a set of small plates connected in parallel. The dielectric for these capacitors is air. Maximum capacitance occurs when the plates are completely overlapped.

160

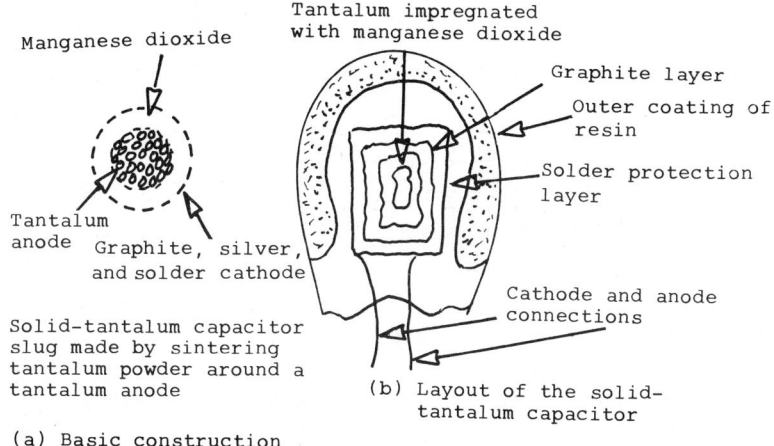

Manganese dioxide

Tantalum impregnated with manganese dioxide

Graphite layer

Outer coating of resin

Solder protection layer

Tantalum anode

Graphite, silver, and solder cathode

Cathode and anode connections

Solid-tantalum capacitor slug made by sintering tantalum powder around a tantalum anode

(b) Layout of the solid-tantalum capacitor

(a) Basic construction

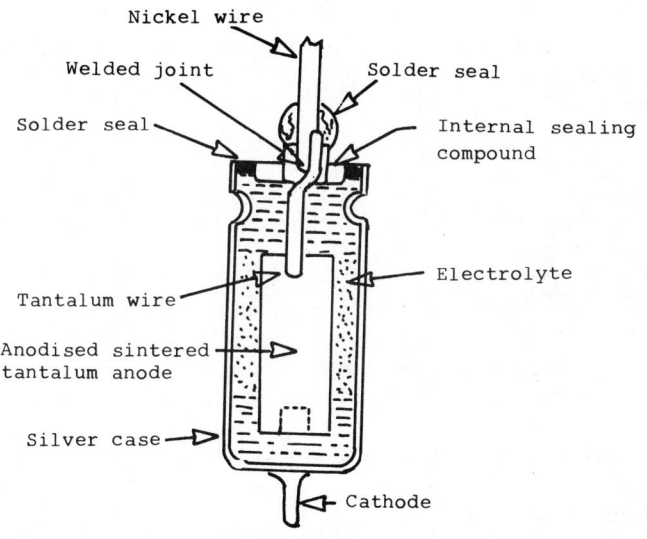

Nickel wire

Welded joint

Solder seal

Solder seal

Internal sealing compound

Tantalum wire

Electrolyte

Anodised sintered tantalum anode

Silver case

Cathode

(c) Layout of wet-sintered tantalum capacitor

Fig. A3.5 Tantalum-capacitor construction

Applications

Typical properties and ranges of applications for the various types of capacitor dielectrics are shown in Tables A3.1 and A3.2.

161

Table A3.1 Capacitor dielectric properties and applications

Material	Working voltage (V) A.C.	D.C.	ϵ_r	Cost per CV*	Size per CV	Tolerance	Capacitance range	Applications
Air (dry)	250 to 1000	750 to 1000	1	High	Large	±10%	1 to 1500pF	Tuning in radio, TV, radar, etc.
Paper metallised	250 to 630	500 to 5000	2.2 to 6	Fair	Small	±20%	0.01 to 100F	General-purpose filters, networks, bypassing, phase shifting at audio and power
film/foil	250 to 630	250 to 630	2.5	Fair	Large	±20%	0.001 to 100F	frequencies; power-factor correction
Mica	6.3 to 630	6.3 to 630	6	Fair	Small	±5%	5 pF to 0.01 μF	High-frequency, high stability, and high-working-voltage capacitors, e.g. standard bridge capacitors etc.
Plastics metallised	250 to 440	750 to 1000	2.3	Low	Small	±20%	0.001 to 100 μF	
film/foil	63 to 500	100 to 1500	2.3	Low	Small	±20%	100pF to 0.47 μF	Superseding paper capacitors for general purpose

* Capacitance × voltage, CV, gives a figure of merit since, in general, the higher voltage a capacitor is required to withstand, the greater the size and cost for a given capacitance.

Table A3.2 Capacitor dielectric properties and applications *(cont'd)*

Material	Working voltage (V) A.C.	D.C.	ϵ_r	Cost per CV	Size per CV	Tolerance	Capacitance range	Applications
Ceramic	63 to 250	63 to 10000	3.5	Low	Small	±10%	5pF to 10pF	Very-high-frequency applications, e.g. tuned circuits and filters; decoupling, bypass, and trimmer capacitors
Electrolytic: tantalum	—	1 to 100	25	High	Very small	±5%	Up to 3500 μF (low voltage)	Large-capacitance smoothing circuits, miniature electronic circuits, coupling capacitors, etc.
aluminium	—	63 to 500	7 to 10	Fair	Very small	±20%	1 to 22000 μF	
Semiconductors (junction diodes and SiO_2)			5 to 10	Low	Very small	±20%	Fractions of pF's to 1000pF	Integrated circuits

TAG tantalum capacitors

The coding system used for tantalum capacitors is shown in fig. A3.6.

Tag Tantalum Capacitors

Capacitance in picofarads				d.c. voltage rating	
Colour	First ring	Second ring	Polarity multiplier	Colour	Voltage
Black		0	x1	White	3
Brown	1	1	x10	Yellow	6.3
Red	2	2		Black	10
Orange	3	3		Green	16
Yellow	4	4		Blue	20
Green	5	5		Grey	25
Blue	6	6		Pink	35
Violet	7	7			
Grey	8	8	x0.01		
White	9	9	x0.1		

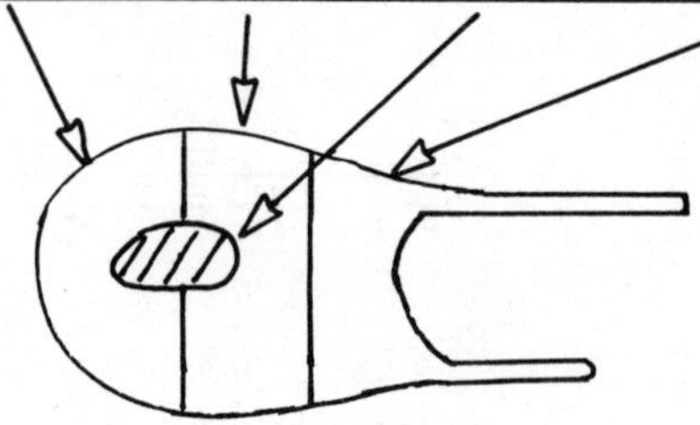

Example:

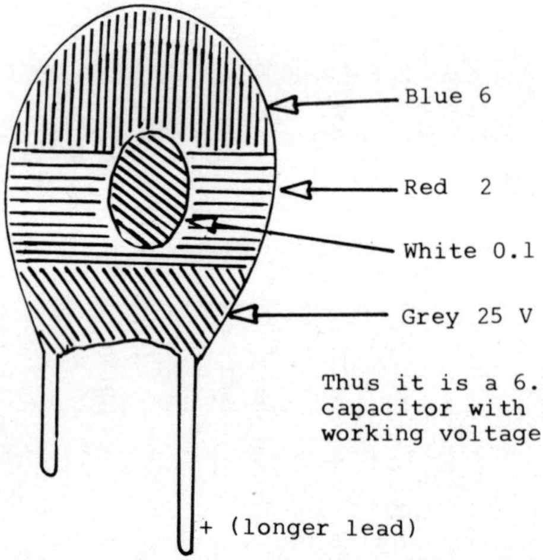

Blue 6

Red 2

White 0.1

Grey 25 V

Thus it is a 6.2 pF capacitor with a d.c. working voltage of 25 V.

+ (longer lead)

Fig. A3.6 Tantalum-capacitor coding

Polycarbonate four-dot and five-dot coding systems

The four-dot and five-dot coding systems used for polycarbonate capacitors are shown in fig. A3.7.

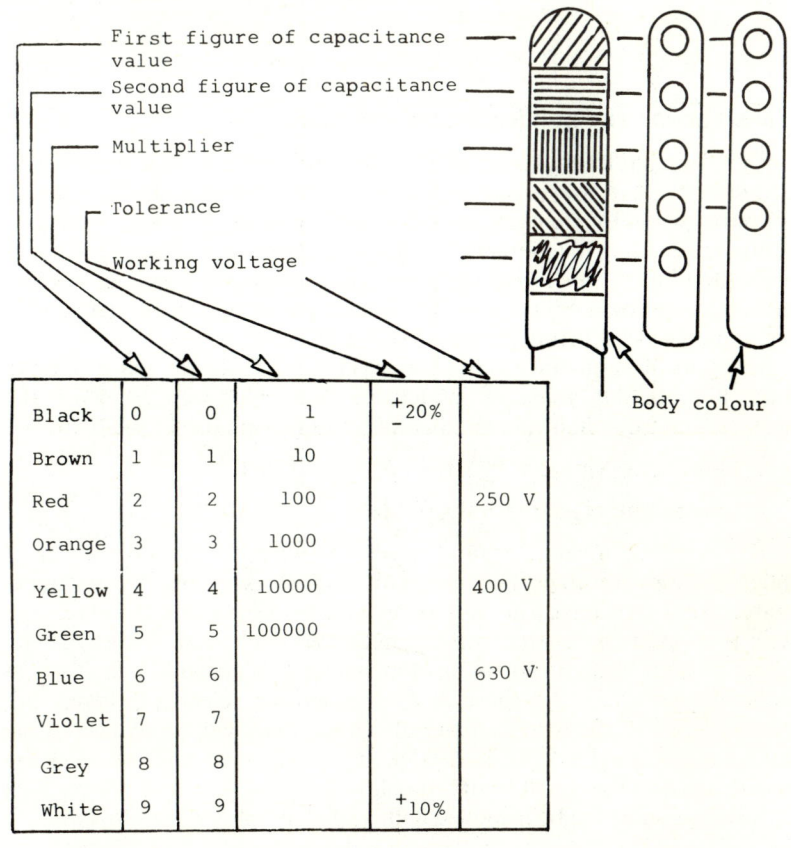

Black	0	0	1	$^+_-20\%$	
Brown	1	1	10		
Red	2	2	100		250 V
Orange	3	3	1000		
Yellow	4	4	10000		400 V
Green	5	5	100000		
Blue	6	6			630 V
Violet	7	7			
Grey	8	8			
White	9	9		$^+_-10\%$	

The resulting value is in picofarads.

Fig. A3.7 Polycarbonate coding systems

Appendix 4: inductors and transformers

Construction and applications of inductors

Basically an inductor consists of a coil of wire wound on to a former. This component then possesses the property of inductance. This is the property whereby changing currents flowing in the turns of a winding produce changing magnetic fields which in turn induce changing voltages into the winding itself (self-inductance) or into other windings in close proximity (mutual inductance). These induced voltages are such that they try to oppose the current changes producing them.

The unit of inductance is the henry (H). Inductors are constructed to have a particular value of inductance, the value depending on the application. For small value inductors, sub-units are used. These are

1 millihenry (mH) = 0.001 H   or 1×10^{-3} H

1 microhenry (μH) = 0.000001 H   or 1×10^{-6} H

The core, or former, around which the wire is wrapped may be of magnetic or non-magnetic material. Inductors having a magnetic-material core will have a greater inductance for a given number of turns of wire than inductors with a non-magnetic core such as air. However, the effect of high frequencies on most magnetic cores prohibits their use at radio frequencies. At these high frequencies, electrical losses are incurred due to the former material. These losses will be reduced if the core is air or if a dust core is used. Some transformers and inductors use ferrite cores. The ferrite core has low losses at r.f. and allows high inductance values to be achieved with a small number of turns of wire.

Air-cored inductors (fixed value)

In fig. A4.1(a), the whole coil is wound into a number of small 'pies' consisting of one or two layers with a small number of turns per layer. This method of construction will reduce the self-capacitance of the coil (self-capacitance has the net effect of reducing the inductance effects and is therefore undesirable). The former consists of a solid rod of poly-ethylene, steatite, or any other low-loss low-permittivity material. The wire is usually enamelled copper, and the pies are connected in series and brought out to connection tags.

Applications   Small-value close-tolerance inductors at r.f. Inductance elements in tuned circuits at broadcast and lower frequencies or as choke coils at r.f. (choke coils are inductors which oppose high-frequency current changes, thus filtering these high frequencies).

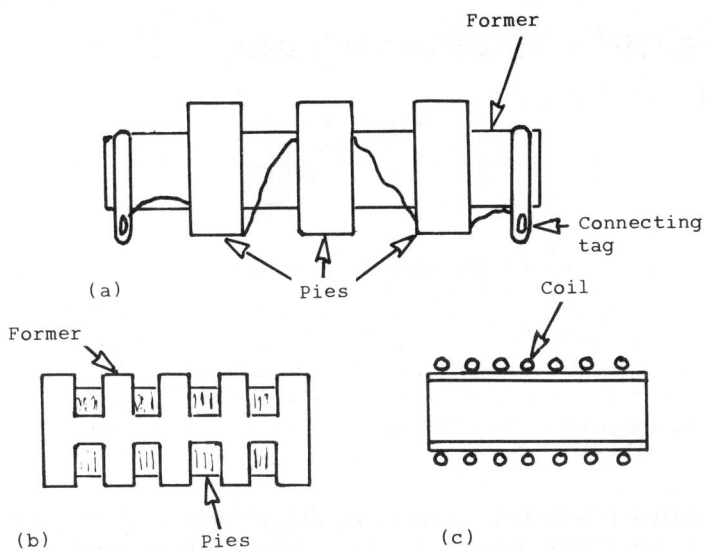

Fig. A4.1 Air-cored inductor construction

Figure A4.1(b) shows another method of construction, using a series of pies wound in narrow slots.

A long single-layer solenoid, as shown in fig. A4.1(c), is sometimes used at r.f.

Magnetic-cored inductors

These are used to smooth out ripple voltages in rectified a.c. supplies.

Apart from the fact that magnetic-cored inductors have only one winding, the basic construction and details of materials are the same as for magnetic-cored transformers, which are discussed later.

Variable inductor (fig.A4.2)

A multi-layer coil of enamelled copper wire is wound on a thin insulating tube of polyethylene or steatite. The ends of the windings are attached to copper terminals which are clamped around the tube. A dust-cored slug is used to adjust the inductance of the winding. The position of this slug relative to the coil can be varied by means of a screw control. Minimum inductance occurs when the slug is totally withdrawn.

Transformers

The basic transformer consists of two or more windings of enamelled copper wire wound on to a former or transformer core. One winding is termed the 'primary' and the other the 'secondary'.

167

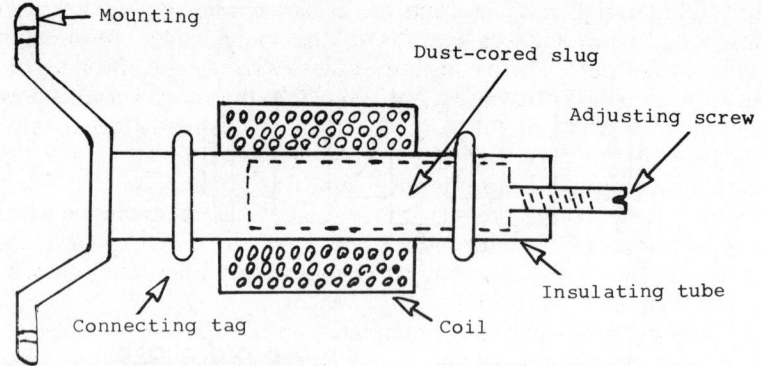

Fig. A4.2 Variable-inductor construction

A current flowing in the primary winding sets up a magnetic field in the same way as an inductor. The core is usually made of a magnetic material, and the magnetic field set up by the primary 'flows' around this to link with the turns of copper wire of the secondary. The magnetic field induces a voltage in the secondary winding, the value of that voltage depending on the number of turns in the primary and secondary windings. Thus a voltage applied across the primary winding will appear either stepped down or stepped up in value across the secondary winding. The voltage has therefore been transformed.

Types of core

Two types of core in common use are made up of thin wafers ('laminations') of magnetic material insulated from each other. The basic shapes of transformers made up using laminations are shown in fig. A4.2.

The reason for using laminated cores instead of a solid piece of metal is to reduce electrical losses (eddy-current losses) which would be excessive

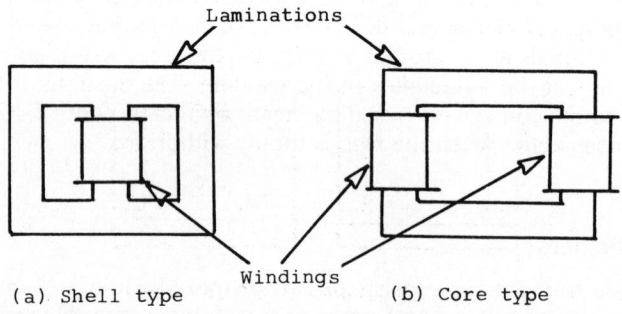

Fig. A4.3 Transformer core types

in a solid material. Eddy currents are caused by the varying flux when using a.c., which induces e.m.f.'s in the metal cores, resulting in circulating currents. These eddy-current losses reduce the efficiency of a transformer. Solid cores have a low resistance, thus large currents flow, resulting in heating of the core. Laminations increase the effective resistance of the core, hence reducing the value of any circulating currents and improving the efficiency of the transformer.

Eddy-current losses increase with the square of the frequency at which the transformer is to be used, and it is found that for radio frequencies the laminations have to be extremely thin in order to reduce the losses. At r.f., some laminations are only 0.05 mm thick.

At very high frequencies the laminations would have to be impractically thin in order to reduce losses, and in these circumstances dust cores or even air cores have to be used.

In dust cores, the magnetic material – iron – is powdered and mixed with an insulating binder. It is then compressed to the core shape and hardened. Ferrite cores are also used at h.f. These consist of oxides of iron, zinc, and manganese, powdered and compressed into shape. They are then fired.

When transformers are laminated, the laminations may take the form of an E and an I shape or a T and a U shape, interleaved as shown in fig. A4.4.

Some laminations may have an I and a U or an L shape as shown in fig. A4.5.

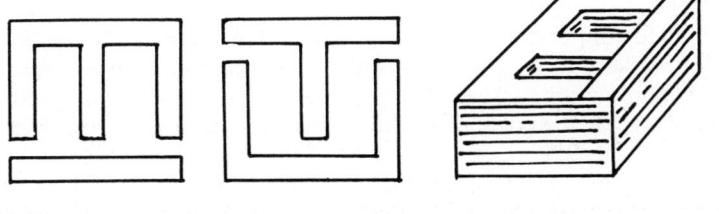

(a) E and I shape   (b) T and U shape   (c) Alternate layers of
                                             E and I then I and E

Fig. A4.4 Types of lamination

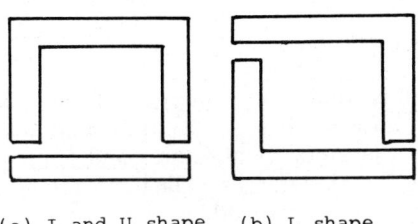

(a) I and U shape   (b) L shape

Fig. A4.5 Further types of lamination

Laminating a transformer is expensive and, to reduce cost, the C-type core may be used. A continuous steel strip is rolled into the shape of the former and then bonded to prevent the coil opening out. The coil is then cut into two to allow the windings to be fitted.

A transformer with a C-type core is shown in fig. A4.6.

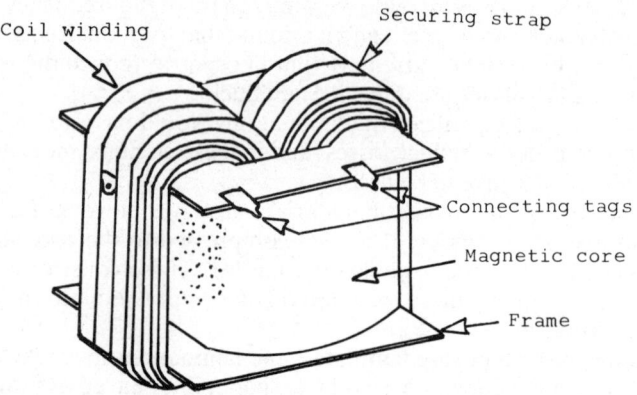

Fig. A4.6 Transformer with C-type core

Windings

Enamelled copper-covered wire is used for the windings. If any thicker insulation were to be used around the wire, the winding would become bulky. The winding is usually wrapped around a coil former and this assembly is then surrounded with the magnetic core in the form of laminations.

Insulating material may be placed over each layer of turns if required, and it is usual to insulate individual windings with insulating tape or Kraft paper.

The winding bobbin, or former, is made of synthetic-resin-impregnated paper built into the required thickness and shape. Thermosetting plastics may be used in some transformers.

Impregnation

Immediately a coil is wound, it is impregnated in a chemically neutral mineral wax or varnish. This gives protection against mechanical damage, prevents the entry of moisture, and improves the dielectric strength of the insulating materials. (Dielectric strength is the ability of the insulation between adjacent sides of the coil to withstand electrical pressure.) Wax is used for low operating temperatures, but for higher temperatures varnish is used.

Air-cored transformers

This type of transformer is nearly always multi-layer and arranged in the form of 'pies'. Each pie consists of a wave-wound layer, and each layer is separated by an air gap. The wave windings cross at an angle, and by this method the self-capacity of the coil is kept to a minimum.

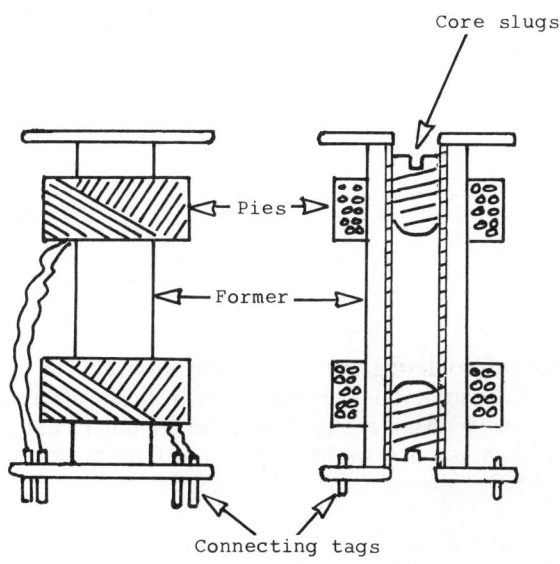

Fig. A4.7 Air-cored transformer construction

The windings are usually wound with Litzendraht wire. This consists of a number of fine strands of wire, each separately insulated and woven . together to form the conductor. Its impedance at r.f. is less than that of a solid-wire conductor.

Tuned air-cored transformers are used in radio-communication transmitters and receivers. Tuning is effected by varying the inductance of the coils. This can be achieved by means of powdered-iron slugs which can be screwed axially in and out of the centre of the coils. Maximum inductance is when the slug is fully inside the winding. A change of inductance of up to 20% is possible with this method.

Appendix 5: soldering and wire-wrap techniques

Introduction

Most faults occurring in modern electronic systems result from the electrical connections between components and the interconnecting wires and printed tracks. It is therefore essential that the service engineer is able to make good connections in the course of repair and replacement of components and circuits.

Soldering

The operation of soft soldering first requires the removal of the oxide film on the materials to be soldered. This is achieved by the flux (for electronic work it is usual practice to use resin-cored solder containing flux). Molten solder is then allowed to run over the prepared surfaces to form a metallic joint. Any surplus flux should then be cleaned away.

Making a good soldered joint

The soldering iron must be clean, well tinned (i.e. the bit must be covered with a layer of solder), and hot. The materials to be soldered should be grease-free and pre-tinned.

The tip of the soldering iron is then placed against the joint to be soldered, and the joint is heated sufficiently to melt the solder. Resin-cored solder may then be applied to ensure that the joints 'run' together. Only enough solder just to cover the actual joint should be used. When sufficient solder has been applied, the solder is removed and then the soldering iron. The joint must not be moved until the solder has solidified.

The function of the soldering iron is to apply heat to the joint, and the iron should not be used to carry solder to the joint. If the iron is used to transfer solder to the joint, some of the active components of the flux may have been used by the time the solder reaches the materials and an imperfectly soldered (or 'dry') joint may result.

A good soldering technique can be achieved only by practice.

Use of heat shunts

Many components used in electronics (particularly semiconductor devices) are damaged by heat. When components are being soldered into a circuit, the heat from the soldering iron must be diverted, or shunted, away from the components. This practice should be used even when temperature-controlled soldering irons are being used.

A heat shunt is any suitable device which can absorb heat. It is placed between the component to be protected and the joint being soldered. In most instances the jaws of a pair of long-nosed pliers or a crocodile clip are quite adequate.

Printed-circuit boards

A printed-circuit board (PCB) consists of copper strips or patterns attached to an insulating board. These strips or patterns replace the separate lengths of wire used in connecting components in a wired circuit.

Care of printed-circuit boards
The copper foil on printed-circuit boards is very thin and is attached to the board with adhesive. Bending or flexing the board may detach the copper from the board or produce hair-line cracks. These cracks will produce intermittent faults on assembly and are extremely difficult to locate.

Soldering printed-circuit boards
Excessive heat from a soldering iron can melt the adhesive and cause the copper track to lift from the insulating board. Heat should be applied only for long enough to melt the solder and secure the component. If a joint is not satisfactory after soldering, let the board cool before attempting to resolder it.

Removal of components from printed-circuit boards

Precautions
a) Power supplies *must* be switched off.
b) Heat shunts should be used.
c) Use the minimum amount of heat necessary.

Removal of components
The desoldering tool is used to melt and remove the solder from the connection. While the solder is molten, the component connecting wire may be gently pulled clear of the circuit.

Practice is necessary to become proficient.

Removal of integrated circuits
Once it has been *firmly* established that an integrated circuit is faulty, it may be removed from the printed-circuit board by cutting the body from the connecting pins and desoldering as above.

If an integrated circuit is to be extracted for testing purposes, the solder must be removed from *each* connecting pin by melting and 'sucking' away, using a 'solder sucker' which is simply a spring-loaded vacuum pump whose nozzle is placed close to the molten solder, the release of the piston creating a vacuum which draws the solder into a cylinder. The integrated circuit can then be pulled clear of the board.

There are many types of desoldering tool available which enable *all* connections on an IC to be de-soldered simultaneously and thus facilitate easy removal of integrated circuits.

Other specialist devices, such as desoldering braid which mops up solder like blotting paper, are quite commonly available and may be suitable in certain applications.

Wire wrapping

As the electronics industry has advanced in technology, the need for faster and more reliable methods of making electrical connections has become important. Also, the miniaturisation of circuits on integrated-circuit chips has required connections to be packed in much higher densities than is possible with soldered connections. To achieve this, a solderless connection using a wire-wrapping technique has been adopted as a standard method of making connections to terminals in high-density electronic equipment.

A wire-wrapped connection is made by coiling the wire around the sharp corners of a terminal under mechanical tension.

A 'Regular' wrapping bit wraps the bare wire around the terminal, fig. A5.1. A 'Modified' bit wraps a portion of insulation around the terminal in addition to the bare wire. This greatly increases the ability to withstand vibration.

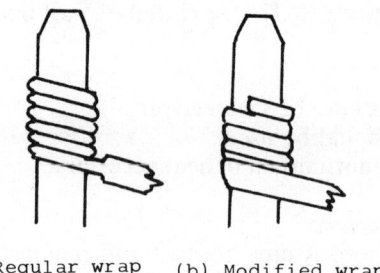

(a) Regular wrap   (b) Modified wrap

Fig. A5.1 Types of wrap

A major advantage of wire wrapping is the ease with which a wire may be removed from a terminal. An unwrap tool is slipped over the terminal and engaged with the first turn of the connection. By rotating the tool, the connection is removed in seconds, without damage to the terminal.

There is a constant surveillance of manufacturing dimensions, and each wrapping bit is subjected to a series of qualification tests. These consist of wrapping groups of wire on various types of test terminal. The wrapped

174

wires are then subjected to a 'strip' test to determine adequate tightness. 'Unwrap' tests are also performed to ensure against an overtight wrap.

Types of wrapping tool
Pneumatic tools are preferred for production work. Electrical mains or battery hand-operated tools are available for service and repair work (fig. A5.2).

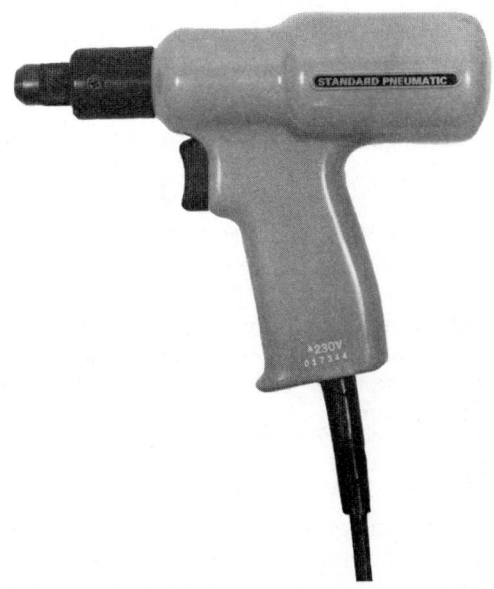

Fig. A5.2 Wire-wrap tool

Wire
Solid wire is used for wire-wrapped connections, copper being the most commonly used.

Advantages of wire wrapping
A wire-wrapped connection is considerably stronger than a soldered one. It is less easily stripped from the terminal and less subject to breakage.

The contact areas of a wire-wrapped connection remain gas-tight when exposed to temperature changes, corrosive atmospheres, humidity, and vibration.

Wire-wrap techniques
1. Press lightly – let the tools do the work. Excessive pressure can lead to overwrapping (fig. A5.3(a)).
2. Keep the tool on the terminal until the wrap is complete. Early removal can result in a spiral wrap or an open-end wrap (fig. A5.3(b)).

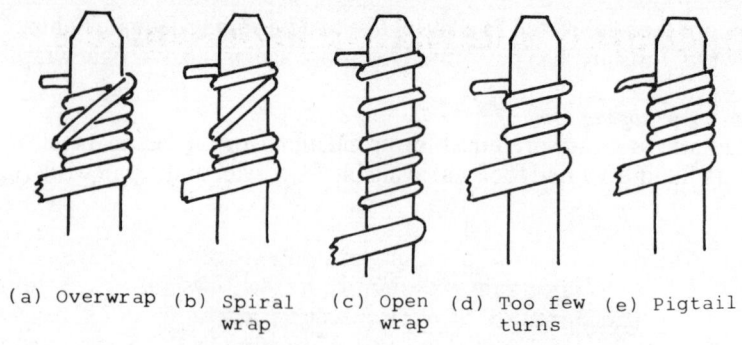

(a) Overwrap (b) Spiral wrap (c) Open wrap (d) Too few turns (e) Pigtail

Fig. A5.3 Wire-wrap techniques

3. Feed the wire correctly into the slot in the bit. Ensure that the stripped end of the wire is pushed in all the way, otherwise insufficient turns may result (fig. A5.3(c)).
4. Check specification tables and use the correct bit and sleeve. Wire wrapping is a precision technique – the wrong bit and sleeve will not do the job. Wrong selection can cause problems such as 'pigtails' (fig. A5.3(d)) and loose wraps.

Crimping

A very useful and commonly used method of connecting cables to connections is to 'crimp' the connections. This is simply a means of providing a tight electrical joint by crushing a suitable sized connector sleeve on to the cable by using a special crimping tool which is essentially a pair of modified pliers.

Many commercial types of crimping system are currently manu-factured and marketed and are too numerous to mention specifically.

Appendix 6: alternative logic symbols

As explained in chapter 6, the symbols used in the text are those of the US MIL-STD 806B and they are in practice the ones most likely to be encountered by the 'users' of electronics, since virtually all data sheets and manufacturers' application notes etc. use them. Also, many practising engineers feel that the shape of these symbols gives them a 'feel' for the logic function.

However, some other logic symbols which may be met are illustrated in fig. A6.1.

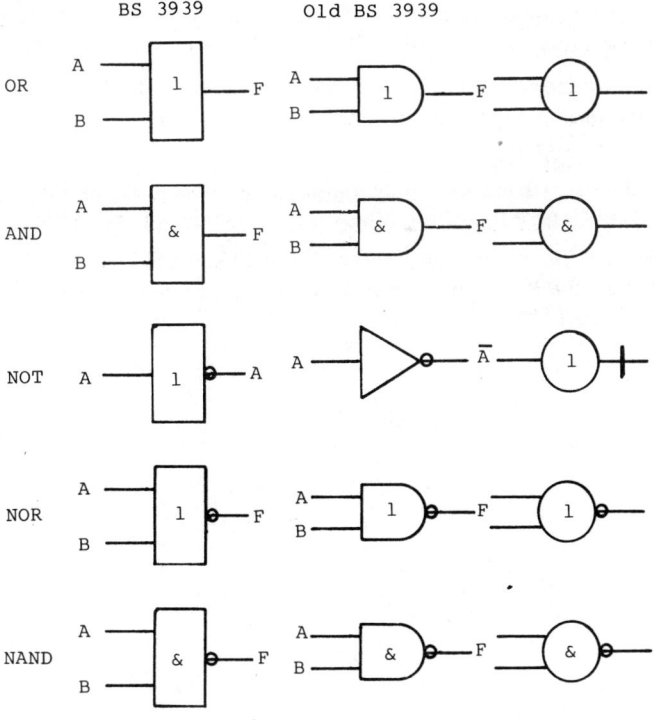

Fig. A6.1 Different logic symbols

References

As mentioned in the preface, in teaching one invariably consults many sources when preparing material for teaching notes, e.g. books, technical journals, data books, etc. The material for this book has been accumulating over a period of about eight years before being put together in what I hope is a form which will be of help in 'spreading the word', and once again I would like to acknowledge my debt to those long-forgotten sources. However, there are particular sources which I have found invaluable in my work and, for those readers who wish to investigate this subject area in greater depth, I have listed them below.

1. *Mullard minibook* series
2. *A programmed book on semiconductor devices*

   Both the above are Mullard Educational Service publications. Unfortunately these excellent little books are no longer in print.

3. *Semiconductor circuit design*, Texas Instruments Ltd
4. *Understanding digital electronics*
5. *Understanding microprocessors*

   The above two are developed and published by Texas Instruments Learning Centre for Radio Shack and are sold by Tandy in the UK.

6. *Understanding microprocessors*, Motorola Semiconductors
7. *Practical digital design using IC's*, Joseph D. Greenfield (Wiley 1977)
8. *SCR manual* (4th edition), General Electric Company of America

Index